MW00609693

Praise for

TREEKEEPERS

"Lauren E. Oakes offers an in-depth portrait of tree planting around the world—its merits and limitations, too. *Treekeepers* brings forests, biodiversity, climate solutions, and global communities into one critical conversation about survival and our collective future."

—Ernest Moniz, former United States secretary of energy

"Oakes knows forests, and in *Treekeepers*, she introduces us to the vibrant global community working to restore forests of all kinds—and ensure our collective survival. An insightful, beautifully reported, and much-needed guide to the hope beyond the tree-planting hype."

—Michelle Nijhuis, author of *Beloved Beasts: Fighting for Life in an Age of Extinction*

"Ecologist Oakes has spent years exploring the potentials and pitfalls of the modern tree-planting movement. The result is *Treekeepers*, a deft and compelling story about the complex realities of growing and maintaining healthy forests, told with just the right blend of hope and warning."

—Thor Hanson, author of *Hurricane Lizards and Plastic Squid*

Also by Lauren E. Oakes

In Search of the Canary Tree: The Story of a Scientist, a Cypress, and a Changing World

TREEKEEPERS

THE RACE FOR A FORESTED FUTURE

LAUREN E. OAKES

BASIC BOOKS
NEW YORK

Basic Books
Hachette Book Group
1290 Avenue of the Americas, New York, NY 10104
www.basicbooks.com

Printed in the United States of America

First Edition: November 2024

Published by Basic Books, an imprint of Hachette Book Group, Inc. The Basic Books name and logo is a registered trademark of the Hachette Book Group.

Print book interior design by Sheryl Kober.
Photographs courtesy of the author unless otherwise stated.

Library of Congress Cataloging-in-Publication Data

Names: Oakes, Lauren, author.
Title: Treekeepers : race for a forested future / Lauren E. Oakes.
Description: First edition. | New York : Basic Books, 2024. | Includes bibliographical references and index.
Identifiers: LCCN 2024005454 | ISBN 9781541603349 (hardcover) | ISBN 9781541603356 (ebook)
Subjects: LCSH: Forest restoration. | Reforestation. | Forest ecology. | Forests and forestry—Climatic factors. | Climate change mitigation.
Classification: LCC SD409 .O225 2024 | DDC 634.9/56—dc23/eng/20240617
LC record available at https://lccn.loc.gov/2024005454

ISBNs: 9781541603349 (hardcover), 9781541603356 (ebook)

LSC-C

Printing 1, 2024

For Calder,
May you always have a sense of wonder and keep that instinct to protect nature. Thank you for marveling with me.

And for Arden,
It took love, science, and a community of caring people to bring you into being—like some other great forests that I now know. I never stopped believing in your possibility.

When I am among the trees,

especially the willows and the honey locust,

equally the beech, the oaks and the pines,

they give off such hints of gladness.

I would—almost say that they save me, and daily.

—Mary Oliver,

from "When I Am Among the Trees"

CONTENTS

AUTHOR'S NOTE

About methods. This is a work of nonfiction, but the story is informed by a blend of extensive reporting and my own experience as a forest ecologist, an expert on nature-based solutions, and a mother helping to raise our next generation in a rapidly changing world. Given the sheer quantity of tree-planting efforts and related research around the planet, I could not include everyone involved. My lens was shaped, in part, by my own history; I am a white woman living in the United States and a scientist who has worked for a conservation organization with restoration initiatives on all plant-covered continents. I've told the story that I am equipped to tell through experience and rigorous reporting, without co-opting the knowledge of Indigenous people whose perspectives on a global reforestation movement I could not fully represent. As a graduate student and then later as a lecturer and adjunct assistant professor at Stanford University, I worked a lot with ecologists, geographers, and climate scientists whose research is highly relevant to this story. Ultimately, whom and what I chose to highlight was informed by 150 interviews that I conducted over four years with researchers, policymakers, corporate leaders, practitioners, and caring citizens who are all working toward a more forested future.

About names and titles. I haven't changed any names. Many scientists earned various degrees and titles such as doctor or professor, but I often left those out for a more informal and conversational narrative. I also didn't want to elevate one form of knowledge more than any other. A person caring for a forest and its ecological community in place over time, for example, offers a depth of knowledge that cannot come from any formal institution.

If I were to identify every person I interviewed or attribute all quotes to identified speakers, there would be far too many characters to follow. I intentionally selected individuals to portray the many ways that people are affected by the state of the world's forests, as well as a diversity of people who are shaping their future. They never called themselves treekeepers, but I thought of them that way.

About travel. I tried to limit travel due to the COVID-19 pandemic and an interest in reducing the carbon footprint of my research. Some of the story draws from travels I took long ago. Other trips were recent and crafted by mixing my firsthand experiences with those I collected from people who live and work in forests near and far.

About reported estimates. Most global targets for forest protection or restoration rely on the metric system. Many scientists tend to do the same for a global standard, using hectares instead of acres. (One hectare is about 2.5 acres.) I have kept estimates in the system that scientists, media, journal articles, and global policies used but often provided equivalencies. The same holds true for tree heights, for example, with most scientists around the world using meters instead of feet. People generally report measurements of carbon or carbon dioxide as *tonnes*, the metric equivalent of a ton, or "short ton," in the imperial system. (One tonne is about 1.1 tons.) In print, there are many references to tons instead of tonnes with respect to carbon, but that usage is almost always driven by the American and British spelling rather than a substantive difference in units. Scientists often distinguish *metric tons* from *tons* for this reason. Unless quoting a print source, I generally use *tonnes* for consistency.

Unfortunately, carbon (C) and carbon dioxide (CO_2) also get jumbled in the tracking of emissions and reductions. Scientists typically use carbon because they are often assessing how it flows through the carbon cycle as mass in different forms, whereas carbon dioxide primarily appears as an emission. Using CO_2 is more intuitive than C for most people who are focused on the gas emitted. Government inventories track CO_2, and most businesses refer to CO_2 because they're similarly focused on the gas if they're trying to make reductions. Some ecologists also use CO_2, and in some cases that makes the most sense, like when estimating emissions from forest fires instead of carbon stored in trees or soil. All this creates some confusion and errors because 1 tonne of carbon is 3.67 tonnes of carbon dioxide. Because my starting point is the carbon that trees can sequester, I generally use C throughout this story and provide CO_2 conversions when possible. I stick to the tendency of reporting the "price of carbon" as dollars per tonne of CO_2, even though more people should really refer to its market value as the "price of carbon dioxide" to avoid yet another misunderstanding.

Lastly, about quotes and conversations. I recorded and transcribed all the interviews that I conducted. I sometimes edited quotes for length and occasionally for clarity. Because I wear two hats as a scientist and journalist, I went back to everyone I quoted to verify their quotes and to offer an opportunity to refine anything further. I felt that effort made for a rigorous yet more comfortable approach, particularly for the scientists I have known in other contexts. In some situations where I was reporting in a remote setting and could not follow up directly, I quoted that person verbatim.

There are also some conversations and scenes I have reconstructed from personal notes and memory. I fact-checked them as a reporter would do. When possible, I went back to other people who were present at the time so that I could describe what happened most accurately.

Many people paused their pressing work in research groups and universities, nurseries, fields and forests, companies, organizations, and governments to speak with me; their openness shaped this story.

PROLOGUE

I had a compass buried somewhere inside my dusty backpack, but a map scrawled on a tattered piece of paper guided me instead. I was twenty-one, on a year's leave from college, chasing a dream of exploring old-growth rainforests in southern Chile. To my surprise, I ended up tracking destruction instead—native forests razed to make way for plantations or consumed by fire. Along the coast, I encountered the remains of ancient trees that were cut in the early colonization of Chile and later covered at high tide by the rising seas of a warming world. On the map I held, drawn for me by a local conservationist I'd met in Valdivia, was a little arrow pointing to *la cruz en el árbol*, the cross in the tree.

Big trees had captured my attention when I was a little girl. I remember spending many afternoons playing under the maple tree in our yard. Later, as a college student, I became curious to learn about various species across the planet, the clearing of forests in American history, and the ongoing degradation of the Amazon. At some point, I'd come across *Fitzroya cupressoides*—the alerce—one of the largest-growing tree species in South America. The species is native to the Andes and the coast of Chile, meaning it naturally occurs in those places. The first descriptions I read of the alerce trees were almost mythic: staggering in size, distinctly fragrant, the startling color of dried blood inside. I wanted to find them.

We had already hitched a ride once that day, then hiked along dirt roads through farmlands. I was traveling with my boyfriend at the time, Stephen. He was a gentle man who went by the nickname Chickie. With his dark hair pulled back into a neat ponytail, he was often mistaken as Chilean, which was occasionally handy. This time my thumb had worked. I ran to the rusted blue pickup truck that pulled over for us and showed the driver my map.

"La cruz en el árbol," I said. "We are looking for the cross in the tree," I clarified in Spanish. The man driving and his partner in the front seat looked perplexed. They weren't from the small village labeled on the paper. After chuckling over our mission, they agreed to give us a ride and see what we might find.

We jumped in the back of the truck and cruised down the bumpy road toward what I hoped was the little dot on the map for this mysterious tree. I didn't know what to expect. The Chilean conservationist who'd told me about it had refrained from sharing much detail. He'd only said that I should go find it, and that was enough for me.

I watched the terrain shift from flat to more rugged and rolling. Then I noticed enormous tree stumps jutting out of the landscape. The astounding size of these remnants, which carried a memory of forests that once were, made for a surreal scene. Each one was far from any other. Their shades of weathered gray contrasted with the verdant landscape that surrounded us. I could not fathom how giant the forest would have been if all the trees were still alive.

We made our way up a hill and round a corner, then Chickie yelled, "Aquí!" *Here!* There was an explosion of cheers from the cab, as the driver and his partner caught sight of the cross too. We hollered back, asking them to stop.

It was not a thriving alerce but another alerce stump, this one bigger than any we had seen yet—some twelve or thirteen feet in diameter. Secured into its core at the heart of the giant was a cross that extended toward the sky, its arms spread wide.

We grabbed our packs, hopped out of the truck, and thanked our new friends before rushing over to the monument. I ran around its base, tracing its circumference like a child spinning around a carousel. The cross was sturdy but old and worn. Maybe it had been there for decades, its origin lost to a history of farmers coming and going to make cropland from forest. I couldn't have climbed on top of the stump if I had tried (and I'm nearly six feet tall). Freshly cut wood from somewhere else in the distant forests was stacked in a long pile at its side.

After standing below the cross in silence for quite some time, we walked around the property and then further down the road past a couple cabins. We found another stump similar in size; a young tree was growing straight out of the chopped one's pith. Maybe someone had planted it, or all the right conditions for another species to establish had just aligned at the alerce's core, its oldest part. I asked a man who was hauling firewood and a couple other people we saw about the cross's origins. No one recalled the story or had any more information to share. My interest seemed to catch them off guard. Maybe the history was to remain a secret to an outsider like me.

Twenty years later, I can still remember the sense of awe and internal conflict that the cross in the tree had stirred in me. It towered over me; yet in proportion to the stump, it appeared like a spindly twig. In the presence of the cross, I'd tried to imagine the tree in its grandeur and the forest, now fields, that had once defined the land and its people. I wondered then, and I still wonder now: When the ancient tree fell, was it a moment of triumph or despair? Who had erected the cross, when, and why? Was it in memory of, and mourning for, the life lost? A testimony to the sacredness of nature? A marker for everything that colonialism destroyed? Was it a tribute to hope or regrowth?

Perhaps it was a sign of what was to come: this idea that trees could be our salvation, or at least part of what might still make the planet more habitable into the future. I'm not referencing a faith-based or religious sort of salvation but an ecological one, based upon the reality of human and plant life as deeply intertwined.

In 2021, the United Nations launched the Decade on Ecosystem Restoration, an effort to prevent, halt, and reverse the degradation of ecosystems worldwide. It was and is a global cry to heal our planet—to restore. That same year, the Declaration on Forests and Land Use was arguably the most important agreement that came out of the UN Climate Change Conference in Glasgow, where leaders came together to address the worsening climate crisis. Over 100 countries committed to ending deforestation by 2030; billions of dollars came from public and private finance to support efforts, including restoration of degraded areas. It was as if governments and corporations across the world were finally recognizing the irreplaceable value of nature—and particularly of forests.

If we had placed markers on the Earth for all the trees in old-growth forests that have been cleared in my lifetime, our plant-covered continents would be studded with tombstones. Still, the forests that remain today continue to suck 20 to 30 percent of the global carbon dioxide emissions that result from human activity out of the atmosphere every year. That's an enormous natural service that trees reliably offer, and it's not their only one.[1]

A new movement has also emerged. At the dawn of the 2020s, many established tree-planting organizations doubled or even tripled the number of trees they planted around the world. The number of organizations and people involved also exploded. Billions of young trees are already in the ground, and billions more are coming. Whether you live in a city and know only an urban forest or you've traveled to see some of the magnificent relics on the far reaches of this planet, the deep draw that people have to trees has in part inspired the almost universal attraction to planting them to combat climate change.

As I watched $1-per-tree marketing campaigns for donations roll out and corporations sit up at the prospect of offsetting emissions in one place with trees in another, the call to write this book became more pressing. I wanted to investigate how our forests are changing. I wanted to explore the big and bold ideas for increasing forest cover and offer a window into the lives of the people who are now shaping the future of forests in an effort to help rescue our own.

I began researching this book in 2018 when I was thinking about these questions: To what extent can forests really save us? And how? As a trained ecologist, a conservationist, and a mother, I was responding to the loss and degradation of forests around the planet and the many impacts, felt already, of a warming world. I wanted a realistic take on what increasing forest cover might achieve for our climate system and much more.

I began by tracking down stories of people and places that have, in recent years or further back in time, reversed a trend of forest loss to one of gain. In my effort to unpack the complicated science surrounding the state of the world's forests today and the prospects for their future, I then turned to the scientists themselves, many of whom I know personally. I wanted to understand their perspectives on the potential of forests to help make a more habitable planet. I visited foresters who were planting trees to get out front of climate change, selecting species and populations that might endure future conditions. I met with seed collectors and employees at start-up companies that are working to scale up forest restoration. Writing this book became a journey to discovery: first, in terms of a global awakening of—dare I say—the sacredness of trees to human survival and how that is manifesting in action, and second, my own as a scientist and a mother, living with the knowledge of colliding environmental crises while also learning from my son's innate appreciation of nature.

This story weaves three narrative threads. The first is an accessible presentation of the scientific facts and controversies about the state of our world's forests and the benefits they offer, the potential of trees to help curtail emissions (spoiler alert: planting trees isn't a silver bullet), and the strategies people are using to restore forests or even create them where they haven't been. The second is my personal journey: as an ecologist and expert in nature-based solutions; as a citizen living out the impacts of persistent environmental degradation in my own backyard; as a mother, discovering natural wonder with my child while feeling deeply concerned about the climate and biodiversity crises. The third focuses on the voices and stories of people who are helping to reverse deforestation trends and sustain forests into the future.

Most dictionary definitions of *restoration* center on returning someone or something to a former condition, place, or position. In this journey I learned that we can never go back; no action today could ever erase what happened or fully bring back what once was. But collectively, people can renovate; they can repair, impart new vigor, revive.

On one of my first reporting trips for this book, I traveled with my husband, Matt, and our then three-year-old son, Calder, to visit his British family in England before heading to Zurich, where I would conduct interviews. We had escaped the persistent cold of late winter in Bozeman, Montana, where we live. In West Sussex, we spent afternoons wandering ancient, sunken footpaths through tree tunnels and fields, while savoring the sweet smells of early spring.

With cousins and grandparents in tow, we headed to Kew Gardens in southwest London, home to one of the largest botanical and fungal collections in the world. On the vast grounds, thousands of specimens live in all their glory; millions of others are preserved as seeds and other plant matter. I watched my boy run past pink magnolia blossoms to the children's area, which is constructed around a huge oak with a walkway through its canopy. A series of circular stones, embedded in the path, led us there. Each stone bore one word carved on its face. A question unfolded as we jumped from stone to stone, reading together in step: What—do—plants—need—to—grow?

Calder paused, took my hand, and looked up at me with his feet still touching the "grow" stone. "What *do* they need to grow, Mama?"

One would think, given what I do, that I would have a clear and immediate answer. Maybe I could just start with a simple biological explanation like "Well, they need sunlight and water and carbon dioxide to make sugar, and that's food and energy to help them grow." Or maybe I could bring his hand to the earth and talk about the soil they need, a place for their roots to take hold and a network of life underground to support the life above. Yet all the complexities of any possible answer to what they need today and will need in his lifetime flooded my mind. Sunlight, water, and carbon dioxide—yes. But so much more. The space to grow. The ability to survive a warming world and what the future climate may hold. They need to be valued and resilient. They require a community of people committed more to their long life than to any use from their death. They need each other to survive, but they also need us humans too, willing to invest in their growth and steward more sustainable relationships with nature into the future.

"They need a lot," I said, as we skipped toward the giant oak. *More than most people conceive,* I thought. As we climbed the stairs into the oak canopy together, I recalled the cross in the tree, that stunning homage to what once was but maybe also to what could be.

If the cross in the tree was a cipher to me when I encountered it many years ago in Chile, today it makes me think of the much-needed rekindling of a healthy relationship between people and forests. Our fates are inextricably linked. Seeds collected and cultivated. Tender seedlings prized and protected. Hands coming together, cupping the roots of the trees for tomorrow.

This book is about the scientists, citizens, and leaders who are seeking to re-create what was lost in the degradation of our world's forests and striving to retain and revive the lingering green. It is a story of the science behind and legitimacy of the global reforestation movement and a critical look at how people might make it better in the years to come. But it is also about learning how to live with a sense of urgency in our warming world and to take actions big and small, while still finding beauty in the present. I believe all the hope tied to this race for a forested future is for our children, if only more people could halt the degradation and support restoration instead.

PART I

VERDANT

The trees are coming into leaf
Like something almost being said

—Philip Larkin, from "The Trees"

1

BASELINE

There was a shift, a tipping point—from nobody knowing about it to everybody knowing about it," Jean-François Bastin told me. "Then it was uncontrollable."

I'd called him from my home in Montana. It was a chilly spring morning, with snow still falling in May. Jean-François was in Brussels, where it was evening. He was sick with a cold but surprisingly willing to talk with me, despite the initial reluctance he'd expressed over email. I understood why he didn't want to reopen old wounds. We were discussing a study he'd led that was published in the esteemed journal *Science* in 2019. It was called "The Global Tree Restoration Potential," and within days of its release, oil companies were announcing tree-planting commitments, and industry proponents had released statements asserting that the most effective way to eliminate carbon is reforestation.

The authors of the study told reporters that from their analysis, Earth could sustain another 1.2 trillion trees. The news headlines focused on planting trees: "How to erase 100 years of carbon emissions? Plant trees—lots of them." *Rolling Stone* later reported that the sensational study was picked up by over 700 media outlets, an atypical situation even for research published in the best journals. It all sounded like the much anticipated and much needed discovery of the silver bullet.

I watched from the sidelines as two camps sprang up almost immediately: those saying, "Let's plant!" and those saying, "It won't work," in reference to this big idea that trees could solve the climate crisis. The camps divided further as planting interest peaked. Governments ramped up their commitments and support. The outdoor clothing brand Timberland launched a campaign to support its tree-planting project, "Plant the Change." Energy and petroleum companies like Total dedicated funding for forests.[1]

Meanwhile, scientists attacked the Global Potential analysis in journal articles as well as the authors themselves on social media platforms. Jean-François was unfairly made to look like a patsy for Big Oil, which is so far from the truth. Some people thought that attempting to discredit him and the science might be the way to correct the misperceptions that tree planting could alleviate the need to reduce fossil fuels. One response in *Science* called the study's estimate of how much carbon these additional trees could sequester "anomalous" and detailed "incorrect" assumptions and calculations. Other ecologists expressed reasonable concerns about how and where planting would occur.

Tom Crowther, the study's senior author, later admitted that he got the communications wrong. He had coordinated the media launch with his research lab's in-house communications team. "It's so naive now," he told me, "but I truly didn't think tree restoration would mean tree planting to the media and the public." Tom had made a splash, a few years prior, with another study that assessed how many trees there are in the world. It estimated 3.04 trillion—about half as many as when human civilization arose.

I remember feeling conflicted at the time the Global Potential study came out and those camps emerged. I thought that perhaps the widening support for more trees could help foster a world that better supports the life within it. The ability of trees to sequester carbon would be only part of that story. I knew there wasn't one quick climate fix, and I was frustrated by those promoting the contrary. I also felt uplifted by the booming interest in tree planting. I still feel that way. Tom told me that the misconstrued messaging was "a terrible mistake" but that it also started so much. It engaged more people. A movement grew.

In December of that year, I ordered holiday cards with a few photos of my little family and received a notice that my purchase had enabled the planting of one tree. "Together with wonderful customers like you," the email read, "we've planted over 1,000,000 trees." It was signed with the hashtag #plantinghope.

"A tree to grow strong with you," I offered in dedication to my infant son. I wondered who would plant that tree, what tree, where, and when. Would it survive? Would it deliver whatever the person planting it or funding it was expecting? I doubt the company's notice was intended to trigger such internal turmoil. But I also knew sustaining a forest or growing a tree is far more complex than planting one.

I wanted to get to the bottom of the debate, to unpack the ways in which growing forests could benefit life into the future and help curtail the climate crisis. I would eventually seek out the people who are attempting to re-create what was lost through many experiences of forest degradation around the world. They are our treekeepers. I would also decipher what restoration has meant in the past in relation to what is still possible today in our warming world.

As a forest ecologist, I had to start with the earliest efforts to map the surface of Earth with satellites and modern technology. "If you cannot map it, you cannot really develop any strategy to conserve it or to monitor it," Jean-François told me. A systematic way of viewing forests around the world enabled more people to track changes. It also helped trigger the dream for more.

An old photograph of Half Dome in Yosemite National Park hangs on the wall of a wood-paneled bedroom in Topanga Canyon in the Santa Monica Mountains. I went to see it and to meet its owner, Virginia Norwood, on a hot, dry summer's day. The photograph is long and narrow, a panorama of the iconic vista. Its thick wooden frame blends into the walls of the room, which has a sweet and musty smell, the kind that makes me think of my grandmother. But there's something strange about the photograph. The colors aren't right. The dome appears ghostly, with its striking face contrasted in dark gray, almost black. The peaks in the distance and the ground exposed beneath the trees are nearly white too. The sky is in shades of gray, not blue. The trees are dense in some places and sparse in others, and they appear like normal trees, except they are red, as if part of a forest in crimson, rust, and cranberry.

It isn't a photograph taken by your typical camera. It was created, instead, from data collected, line by line, by sensors on a multispectral scanner system (MSS) that would soon orbit Earth for the first time ever. A dashed box outlines the word "HUGHES" in the far-right corner beneath the image—for Hughes Aircraft Company. The light reflected by that landscape was captured from 2.5 miles away, from the ground, during a trial run before the MSS went to space.

In reality, those red forests were just as real as any green ones; we're just used to seeing trees as green with our own two eyes. The sensor captures light of different wavelengths as separate bits of data, and then scientists assign them colors. The MSS can register more light—such as infrared—than is visible to humans.

"Upper management wanted pictures," Virginia said, seeming a bit disgruntled at the memory of the prelaunch testing requirements. She and her colleagues had run a lot of lab tests while refining the prototype. "Since the test crew were Californians, they had to go to Yosemite," she chuckled.

"You can probably see that she didn't think additional testing was necessary," her daughter, Naomi (or N^2, as Virginia preferred), chimed in. Nevertheless, Virginia outfitted the crew with the engineering model of the scanner, which was mounted to a turntable in the back of a truck, and then sent them to acquire pretty images. They would show what the invention could do before the 105-pound flight model was launched into space that same year: 1972.

Virginia was ninety-five years old when I visited her. Every morning she climbed a rickety step ladder to pour seed into bird feeders for the Oregon Juncos, House Finches, and Spotted Towhees that darted beneath the branches of the giant coastal oaks that shaded her home. Naomi, her family, and those who have known her well call her "Ginger." Only recently had she become more widely known as the "Mother of Landsat," the first satellite launched with the express intent to study and monitor the surface of the Earth.

Virginia's high school guidance teacher, when seeing her aptitude test results, had suggested she become a librarian. She applied to the

Massachusetts Institute of Technology instead, and when Virginia began her studies there in 1944, she was one of only about a dozen women in her class. She finished just three years later, and it didn't take long for her inventions to go to the moon. When she joined the antenna lab at Hughes Aircraft, she was the only woman among some 2,700 men in research and development. By 1966, when Surveyor 1, a robotic spacecraft, landed on the moon in the first fully controlled soft lunar landing, it transmitted images back to Earth on Virginia's transmitter. She gave us the ability to see the moon up close and then, soon thereafter, to rediscover and monitor our own planet.[2]

Orbiting Earth, the satellite could scan the entire planet every eighteen days. It was called the Earth Resources Technology Satellite (ERTS) before being renamed, more aptly, Landsat. Among many other things, it began producing our first global view of the world's forests from space.

The ability to view infrared reveals more contrast between vegetation, burn scars, and other features of land cover. An image captured by the sensors may not appear as one would expect, but it can document unique attributes of the land that would not be visible otherwise.

Virginia's scanner could capture more information than the human eye, but even more remarkably, it did so systematically. When launched into space, it followed 251 precise paths around the planet to acquire complete coverage, over and over again. It collected the light data and processed it into pixels, each one corresponding to a patch of Earth about the size of two regulation-size ice rinks.[3]

Virginia, Naomi, and I made our way into the living room and sat down on a couple of couches across from one another. Naomi passed me the final report from the testing mission, which documented test runs of various scenes and subjects under different light and cloud conditions. I thumbed through and stopped on a page with eight different images of vegetation up close. The labels beneath each one revealed bell peppers, grass, strawberry, brussels sprouts, dichondra, broccoli, and lettuce. Presented in a grid of rectangles, the images looked like natural color photographs, slightly faded.

"But the data!" Virginia said. "We used the plants to show that the data the sensors collected from each one was different than the others. There was a unique spectral signature for each plant."

A spectral signature is an essential element of modern-day *remote sensing*, the process of detecting and monitoring the physical characteristics of a given area (a forest, in my case) by measuring its reflected and emitted radiation at a distance. There it was in its earliest form, showing that when the scanner would later orbit Earth, it could discern vegetation types and much more. A forest attacked by beetles or singed by fire would have a different signature than a healthy one. The data could reveal bare ground exposed by roads cut and carved into intact, verdant forests.

When Virginia was working to develop the MSS, she had met extensively with prospective users of the data. There were all sorts of desired uses, from assessing agricultural crop production to tracking water pollution. Tracking forest health was just one potential application among many, including military uses and national security. Yet how that scanner revolutionized forest monitoring is what had brought me to Virginia.

When ERTS was launched into space on a rocket on July 23, 1972, it was the first of what would become the Landsat Program, a series of satellites that rolled out with improvements over the years and decades that followed. Virginia had watched the launch with a crowd of observers from bleachers, along with her partner and her two boys. The next day the *New York Times* proclaimed that "a new era of Earth exploration" had begun, under the headline "An Earth-Exploring Satellite Is Orbited." After a first successful year in orbit, the satellite was deemed a second "giant leap for mankind" because of its potential to improve understanding of environmental issues.[4]

I asked Virginia what she was thinking about when she witnessed the rocket blasting off, because to me it marked the start of everything people now know about the state of the world's forests.

"I was just hoping that everything worked well," she said. "It did, surprisingly."

Virginia had been so focused on the intricacies of the invention and the various challenges she needed to resolve along the way. She wasn't thinking about how that satellite would change the world. She didn't know it back then; nor did she really acknowledge it years later in her living room with me. But, from my perspective, Virginia was the first of the modern-day treekeepers.

Within days of the launch, the MSS collected imagery of an 81,000-acre fire burning in isolated, central Alaska. In a single image, scientists and resource managers could assess the extent of fire damage, for the first time ever, while a forest was still burning. The data that Landsat 1 collected inspired immediate changes to practices in the fields of cartography and geography. Country boundaries were redrawn. Off the coast of Canada, an uninhabited island was discovered and named "Landsat Island." For self-defense, the Russians had been creating false maps with cities located on the wrong side of rivers, for example, so that foreigners wouldn't know where they were. The aim was to confuse the enemy in the event of an invasion or simply to lead anyone too curious astray.[5]

"Suddenly, everyone in the world had a better idea of what was happening in Russia than the Russians ever had themselves," John Townshend, an early user of the imagery for forest monitoring and a remote-sensing specialist, later told me. "You couldn't overestimate what an incredible surprise it was."

The National Aeronautics and Space Administration (NASA) offered grants for scientists to put the data to all kinds of uses. Not long after the launch of Landsat 1, forest managers and scientists started using the imagery—along with images produced from data collected by subsequent satellites—to assess forest health and degradation in priority areas. Computers enabled sophisticated analyses. Sometimes researchers would also sit around light tables with the prints, characterizing what they saw in one plot of Earth after another. The burgeoning field of forest monitoring was largely dominated by scientists in developed countries who could access the competitive grants or pay for the imagery and also had the resources to analyze the data.

Later, when thumbing through a book about Landsat history, I came across a grid of four images acquired in 1975, 1986, 1992, and 2001 over Rondônia, Brazil. They were color images from Landsat sensors, bright green as one would expect for the dense Amazon. They brought me back to one of my first remote-sensing classes as a graduate student at Stanford, when we looked at satellite imagery documenting land-use change around the planet—urban sprawl, agricultural production, recently cleared forests. The Rondônia images revealed what I had learned was called a

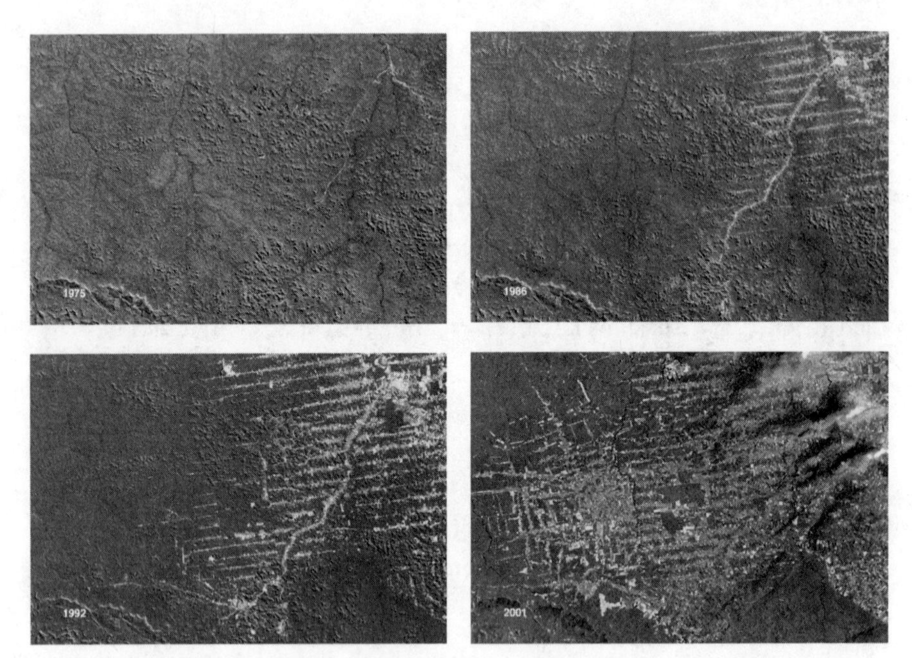

Landsat images showing deforestation patterns (crosswise from *top left* to *bottom right*) in 1975, 1986, 1991, and 2001, respectively, in Rondônia, Brazil.

Credit: NASA/USGS Landsat

herringbone pattern. A new road appears as a thin line of cleared terrain (the spine); it's followed by the appearance of clearings on either side of the road (the ribs running perpendicularly). Over the years, the clearings then expand from the road outward and into the forest. I could see the pattern in the series: first, the intact forest; second, the main road running north to south; third, the offshoots spanning east to west; last, large patches of forest cleared—a fish bone spine with the ribs exposed and the flesh in between getting picked away.[6]

The multispectral bands, as well as the consistency and continuity of the data collection with every orbit of Earth, were breakthroughs. But the time series became even more valuable, as Landsats 2, 3, 4, 5, 7, 8, and 9 entered orbit in the decades that followed, along with thousands of other satellites.* A series can reveal change between years or months or even day to day.

* Landsat 6 failed before reaching its intended orbit. By 2022, around the fiftieth anniversary of Landsat, there were approximately 6,500 known satellites in orbit, observing the Earth's surface from a variety of altitudes and angles. The US Geological Survey (USGS) Earth Resources Observation and

"You can't map a forest with a single image with confidence because there's a lot of other vegetation that can look like forest at any given instant," Matt Hansen told me over a long-awaited discussion (on my end, that is) about the state of the world's forests. A remote-sensing scientist who is famous among those obsessed with forest monitoring, he is known for the "Hansen Dataset," which I'll get to soon. It revolutionized global forest monitoring and put the ability to track forest change—anywhere on the planet—into the palm of anyone's hand.

"You need multiple looks, and then you need multiple looks to reinforce the signal. You want to see the forest repeatedly. Even just identifying forests as a thing, you need a time series, and of course you need time series to say it has been changed. The time series is everything."

The time series is critical, but so is the baseline, the first reference point, for understanding what was present before in relation to what is there now.

Around the same time that I began researching this book in 2018, I gave birth to my son, Calder. Some hours after his arrival, my nurse, Mary, came into my room in the middle of the night. She offered me a few pain pills in a plastic cup. I took them from her, gulped them down with a sip of cold water, and then asked if she could bring them next time without the plastic waste.

Caught off guard, she agreed, then inquired, "What do you do?"

"I'm an ecologist; well, an interdisciplinary environmental scientist, really. I work mainly on climate change; you know, helping people figure out how to cope with the changing conditions—like drought or fire here in Montana." She leaned in a bit closer, seeming perplexed but also intrigued.

"And you just had a baby," Mary stated. She paused, then added, "So, you must have hope."

Science (EROS) Center's Landsat archive contained over ten million images. The image archive has continued to expand rapidly. (EROS Center, "Landsat: Celebrating 50 Years [Extended Edition]," USGS, 2022, https://www.usgs.gov/media/videos/landsat-celebrating-50-years-extended-edition; Landsat Missions, "Landsat Archive Adds Its 10 Millionth Image," USGS, December 15, 2021, https://www.usgs.gov/landsat-missions/news/landsat-archive-adds-its-10-millionth-image.)

The pain from the birth was still pulsing through my body, and the adrenaline, veiling the exhaustion from two days of labor, had begun to wane.

"Yes, I chose to bring a child into this world," I said, now feeling a bit defensive and stunned that I was having this conversation hours after becoming a mother. "But he will come to know a far different world than the one that you and I have known," I added.

Over the years, as his tiny hands have grown bigger in the gentle grip of my own, I've internalized the concept of an ecological baseline through my boy. What I have witnessed as loss and environmental degradation is but a starting point for my son, born seeing beauty in yet another beginning. We are all born into our own "normal." We have different starting points for comparison. Change gets lost in perception, across generations and time. His normal includes smoke from wildfires seeping into our Bozeman home, water-use restrictions, and farms producing vegetables and grain where forests previously flourished.

The innate reverence for nature and trees, specifically, was there for him too. I remember the tears that fell when he'd learned, at age three, that we'd have to remove a dying Douglas fir in our yard.

If we forget, or overlook, or just don't know what was here before, how can we ever want it back? Or endeavor to create it anew?

If 1972 marked the birth of fine-scale monitoring of the Earth, 2009 brought another breakthrough. A new distribution policy, passed in 2008, opened the entire Landsat archive to the world for free. Anyone could look back in time.

"We used to say that we use the data we can afford, not the data we need," Matt Hansen told me. We were meeting on the morning that the first images of galaxies taken by the James Webb Telescope were released in 2022, expanding the modern gaze outward to the great beyond. "When they opened the Landsat archive, it was like, 'What data do we need? Well, I guess all of it!'"

The internet and Google catapulted humanity into a new era of accessible Earth observation and analysis. In June 2005, Google Earth launched, bringing a full view of the planet to personal computers around the world. Anyone could travel virtually for a bird's-eye view of the Amazon forests, a city in Europe, or their own backyard. Five years later, Google brought the ability to analyze the imagery to everyone with Google Earth Engine (GEE), a planetary-scale platform that made studying changes far easier than it had ever been.

Before the engine went public, Matt had come together with software engineer Rebecca Moore, founder of the project that would become GEE, in a timely encounter on the grounds of the National Institute for Space Research in São José dos Campos, Brazil. By the time they'd met, Matt already had a vision for the first-ever global analysis of forest cover using Landsat data. But the *how* was still unresolved.

"I learned of her plans for GEE, and we took it from there," he told me.

The mission was to take all the imagery from Landsat 7, map the extent of forests in 2000, and assess any changes since. It would be a massive undertaking to learn from so many images that had been acquired long after Virginia's MSS first took flight.

"At that time, we couldn't do the data processing ourselves at that global scale. Our algorithm worked, but we couldn't run it across all the images that were available. We had the image-processing and characterization expertise, and Rebecca's team had the computing-accuracy expertise," Matt recalled. "So, we put that together, and we ran the planet, and it was awesome."

It took one million hours of computing time on 10,000 computers processing 654,178 Landsat scenes to characterize the forest cover on the planet in 2000 and the subsequent loss and gain through 2012.[7]

Forest loss, as they tracked it, meant some sort of disturbance had eliminated the trees or left them standing dead. Between one image and another, or gradually over time, the forest cover disappeared from a given pixel of data. Sure, there were other places experiencing a gain in forest—where natural forests regenerated or forestry practices took hold, greening the landscape in neat rows for another kind of crop. But in the tropics, forest loss was still increasing year to year. Analyses of the Hansen Dataset showed where the remaining forests were (in 2013), but even more revealing were the rates of loss documented around the planet. According to several targeted studies and monitoring efforts, deforestation in Brazil had slowed. However, Matt and his colleagues showed that it was increasing in other countries, such as Indonesia, Malaysia, Bolivia, Zambia, and Angola.[8]

In 2018, scientists came forth with a denser times series of forest change by making use of imagery with a higher temporal resolution, meaning less time between scenes captured at the same location than what Landsat offered. Due to bigger pixels, the images had coarser detail than Matt's images, but this global effort took the baseline back to 1982.[9]

"We had satellite data every day for thirty-five years," Xiaopeng Song, the study's lead author, told me when I tracked him down. (Matt Hansen was a coauthor.) In his office at Texas Tech University, Xiaopeng's had a NASA calendar opened to the month that happened to portray Virginia Norwood.

Xiaopeng was careful in conversation to refer to "tree canopy cover," or "tree cover" for short, instead of forests or forest cover because he could focus on the trees themselves and not on what kind of forest they constituted together. Across the entire planet, when averaging out changes everywhere, Xiaopeng and his colleagues could see an increase in tree cover. The overall gain in canopy that the study revealed was "a result of net loss in the tropics being outweighed by a net gain in the subtropical, temperate, and boreal" regions. Forest destruction in some places was offset by creation in others: where natural regeneration could occur, where plantations were growing, where trees were planted in places they hadn't been before.

"We could see the effect in China," Xiaopeng told me; plantations had created a dramatic increase in tree cover.

Prior to the era of remote forest monitoring, the Food and Agriculture Organization (FAO) of the United Nations had been conducting the most comprehensive global effort. Every five years, the organization would compile data submitted by countries from their national forest inventories—data collected on the ground and other observations. The first survey was published in 1948; 101 countries responded, representing about 65 percent of the world's forests. FAO still publishes its Global Forest Resource Assessment, although participation and the extent of coverage have varied over the years, self-reported data have inaccuracies, and differing definitions of *forest* among countries has also caused problems. Some experts call the early FAO figures from the country-level data hopeless for these reasons. Nevertheless, the most recent report (2020), compiled from inventories conducted in 236 countries, found that forests covered 4.06 billion hectares, or approximately 31 percent of the world's land area; that estimate of the total is close to the one first reported in 1948. I was surprised when I made the comparisons. But the result also made sense, considering Matt Hansen's story of forest loss and Xiaopeng Song's version of tree-cover gain.[10]

Jean-François Bastin told me there are good reasons to question what countries report and to criticize satellite data. Tree or forest cover seen from a satellite can only reveal so much. In a story of one pixel shifting from loss to gain, what those forests are—under the canopy and inside the green—may have changed. Plot data and ground-based inventories can describe a local forest in fine detail, but the fieldwork required means there's a limit to that coverage too. He said that people often like to communicate one global estimate or results from a particular study, but that has limitations as well. "If you put everything together, you can get a good sense of what you really have."

Felix Finkbeiner was nine years old and living in Bavaria when a school assignment on Wangari Maathai touched him in a way that I imagined to be instinctual, like Calder's reactions at Kew Gardens and in our backyard. Maathai was the first African woman to be awarded the Nobel Peace Prize. Among many other pursuits and accomplishments, she'd started a movement in Kenya to counter deforestation, to plant, and to restore. Felix challenged his classmates to plant trees, and then he went after a bigger vision. He wanted every country to plant a million trees. He founded Plant-for-the-Planet, an organization that would help them do so. Years later, with his rimless glasses and emphatic tone, like the climate activist Greta Thunberg, he was speaking before the United Nations at age thirteen and advocating for a trillion trees. That was in 2011, long before the analysis for "The Global Tree Restoration Potential" study was even an idea. Somewhere along his journey, Felix must have seen forests as more than pixels of green in a global portrait. He wanted to know how many *trees* there are on Earth; not just where "forest" is.

Years later, Tom Crowther was a postdoctoral fellow at Yale and living with a housemate who had met Felix in Germany and had helped Plant-for-the-Planet. Tom was studying fungi and didn't know a lot about working with satellite data at the time. His housemate was trying to learn how many trees still existed but couldn't find any reliable estimates. The search piqued Tom's interest.

"I thought it sounded like a fun challenge," Tom told me. "So, I went for it. I thought, 'Why can't I study the global forest system?'"

In 2015, their result of three trillion trees made headlines, of course. In the analysis, Tom and the other scientists used nearly half a million tree-density measurements collected on the ground and across every continent except Antarctica. Yet so much that transpired in forest monitoring since Virginia's scanner first launched had also made the tree count possible. Four years later, and by then a world-renowned ecologist, Tom, along with Jean-François and their colleagues, revealed how many *more* trees the Earth *could* support. They also offered another number that many people would fill with hope: how much carbon those trees could sequester.[11]

2
THE POTENTIAL

"Out of nowhere, I got invited to a conference in Beijing to present my work as a keynote speaker," Jean-François Bastin told to me when I asked how he and Tom Crowther first met. Jean-François had very politely declined to speak with me when I'd initially contacted him about the famous Global Potential study. He'd said that he wasn't comfortable with all the buzz it had generated; he preferred to leave the communications to the Crowther Lab, the research group that Tom had built at ETH-Zurich after his years at Yale. I wrote back begging for a chance, letting Jean-François know that the science was solid; the clickbait media and controversy couldn't be erased, but people do move on; his perspective was important and still unheard. I'd scoured the internet for his story but found little in comparison to the hundreds of news articles that quoted or profiled Tom.

"Make sure you talk to Bastin," Tom had told me when we met in Zurich.

In my digging, I'd discovered that Jean-François's account on Twitter no longer existed. (This was long before Elon Musk took over and users dropped for other reasons.) Traces of the controversial backlash were still there: "No need to stop imperialism and resource extraction. Just plant trees and it's magically fixed"; "The audacious effort to reforest the planet"; "Mainstream reporting of the 'Plant Trees to Fix Climate' (Bastin) study is **EXTREMELY** misleading!" Articles included a quote from Jean-François, who had apparently said, "And this is a beautiful thing, just to think that in order to fight climate change what you have to do is to plant trees, and you can do that everywhere." The quote was from a publicity video that the Crowther Lab had released around the 2019 publication of "The Global Tree Restoration Potential." But when I went looking for that specific clip, the links were broken. It too had disappeared.[1]

When I searched again, later in 2021, I found the video on YouTube. It was short and sweet, but it must have been edited and uploaded as another version at some point. That quote from Jean-François had been removed. It was a slick production with dynamic illustrations and a fun base beat to emphasize some highlights: they'd used high-resolution satellite photos from 78,000 points—each one covering one hectare—around the planet. With machine learning and artificial intelligence aiding analysis of the "big data," they'd tied those points to more information about environmental characteristics. The study relied on a three-step process: (1) identify where trees might naturally exist in protected areas, (2) generate a global map of where trees could *potentially* exist in the absence of humans, and (3) subtract away the areas that we cannot restore, like urban and agricultural land. That left behind a map of the great reforestation potential. "The area for extra tree-restoration potential is far greater than we could have ever imagined," Tom reported in the video. "Ultimately, these models reveal for the first time that we have a fantastic opportunity to manage and restore these ecosystems in the fight against climate change."[2]

"I've always found maps quite beautiful," Jean-François told me during my first conversation with him. I'd promised him that I wasn't looking for flashy headlines or a brief quote for an 800-word news clip; I wanted to hear his story. He agreed to talk and then spent hours with me.

As graduate students eager to learn remote sensing, we had both studied under Eric Lambin, a Belgian geographer who, along with scientist and author Jared Diamond, won the Blue Planet Prize in 2019 for his life's work on land use and the many factors that drive people to clear forests or to plant them again. Jean-François had worked with Eric in Belgium, so we never overlapped in person. He'd been trained as a forest engineer, applying engineering principles to maintain trees, soil, water, and other natural resources within forest ecosystems. Using satellite data for mapping became another attractive tool for him.

Despite his own surprise at being asked to speak in Beijing in September 2017, the invitation made sense to me. Experts from many countries were coming together for an inaugural conference on the use of big data to understand forest ecosystems. As a PhD student, Jean-François had spent months at a time living in the core of the Congo Basin, where he measured trees to assess the carbon stock. To estimate the amount of carbon stored, scientists like Jean-François commonly use estimates of above-ground biomass, the standing dry mass of live or dead matter from a tree. The quantity stored differs between trees and forest types. Assessing that variation at the tree-level often starts with very physically challenging work.

He'd gone to the heart of one of the most diverse, carbon-rich forests on the planet—a place also experiencing a huge impact from slash-and-burn clearing and illegal logging. There, Jean-François developed a better understanding of how much carbon storage can vary inside a diverse forest. Forested areas just 500 meters apart can have very different species compositions, structures, and wood volumes—and, therefore, amounts of carbon as well. He doggedly collected tree measurements and survived mainly off canned tuna. But his months there in a remote place where few other forest datasets exist even today also gave him an understanding of how complex natural forests are. On the ground, he could see and record so much more detail than any satellite could.

"I worked in an area about thirty kilometers by thirty kilometers. To the south, there were forests habituated by bonobos. To the north, they were inhabited by elephants. There were trees pushed to the right and to the left to make room for elephants. Without seeing them, you could feel they have been there—that they are there." Some trees, such as the *Gilbertiodendron dewevrei*, had monumental trunks and grew to over 100 feet tall. At times, the leaf litter came up to his knees. Jean-François said it was like trekking through deep snow powder. From the forest floor, he could see how the trees were intertwined, their branches reaching across to each other and creating clouds of green.

"You can see that from the satellite," he told me. "But you can't single out a tree below." That's what he did on the ground; in each plot, Jean-François and his team would measure some 400 trees. He measured over 10,000 trees and recorded 250 species to estimate the carbon stored.

Later, as a researcher at the Food and Agriculture Organization (FAO) of the United Nations, he led an effort to compile and analyze satellite imagery from more than 200,000 plots around the planet. The goal was to provide a global estimate of forests in drylands, which are low-rainfall areas often characterized more by grasses and shrubs than by trees.

"It's a bit like you have a map, but the map is showing you that you have nothing," he explained, highlighting the occasional inaccuracy of some global maps to show where trees really are. People living in adjacent communities or working for a local nongovernmental organization (NGO) or government might know about a forest in one location. But if that forest isn't showing up in any official global maps, it's difficult to convince the national government to help protect or manage it. Pixels may reveal only part of the real story.

Yet that study, the most intensive of its kind to date using high-resolution satellite imagery from dryland ecosystems across the world, showed far more forested areas than previously documented. An impressive work for the thirty-one-year-old researcher, the drylands forest study had landed in *Science* in 2017 and then brought Jean-François to Beijing. After he had presented, Tom Crowther approached him a day or two later. He said he was building a research lab and wanted Jean-François on the team.[3]

"I had never had an interaction like that," Jean-François said. "I didn't know if he was joking because he was also super young. He was younger than me. It took about five days to think it through. But then I thought, 'Why not?'"

"You can do whatever you'd like. What would you like to do?" Tom had said.

"I told him I wanted to map the tree-carrying capacity of the planet to see if the big pledges by NGOs and others about restoration were doable or not. I told him that I could do it with the data I had been using at FAO." Jean-François already had a nice systematic sampling of the planet. By the time he and Tom met, the Bonn Challenge, a target launched in 2011 by the government of Germany and the International Union for Conservation of Nature (IUCN), had been well underway to restore 150 million hectares of degraded and deforested landscapes by 2020 and 350 million hectares by 2030. There were other aspirational commitments too, like the New

York Declaration on Forests, a 2014 pledge to halve deforestation by 2020, to end it by 2030, and to restore hundreds of millions of acres of degraded land. However, there wasn't the booming interest in tree planting that would come from what Tom, Jean-François, and their colleagues deliver later.

Tom had replied, "This is exactly what I want to do too! If you had told me you wanted to do something else," he continued, "I would have told you I want you to do this." Jean-François was grinning, recalling the energy of that moment and their perfect alignment.

He returned to Rome, where he was living and working at the time. He put everything in place to leave and said good-bye to his friends and colleagues at FAO. By the time he arrived in Zurich just a few months later, Jean-François already had some preliminary results.

"If we removed humans from the planet, what would the planet look like with respect to forests?" he recounted to me in his Belgian accent. "That was the idea for the paper." If they could compare that ideal, unrealistic situation with the current situation, they could assess the differences. Then they could see where there might be room for restoration, "while respecting what we would have without the human species."

He called it "an attempt to build the planet without us." The data crunching went quite quickly from that point forward.

"The Global Tree Restoration Potential" was published in *Science* on July 5, 2019. Jean-François told me his original title was "The Tree-Carrying Capacity of the Planet," an ode to what forests might be without us. Later, with some regret given all that unfolded, Tom admitted he wished it had been "The Natural Tree-Regeneration Potential" or something along the lines of "The Natural Recovery Potential." This all probably sounds like semantics, but the words really did matter and still do—especially when the world is seemingly on fire; when floods and droughts are hitting hard; when fossil fuel emissions and forest clearing persist; when anyone even slightly concerned about climate change wants the answer to be clear and simple, like, "Here's the one magical solution! Let's do it!"[4]

The researchers found "room for an extra 0.9 billion hectares of canopy cover" on Earth, a total area about the size of the continental United

States. They calculated that such an increase in canopy could store about 200 gigatonnes of carbon, roughly equivalent to two-thirds of the total "carbon burden" in the atmosphere as a result of human activity.* Later, they clarified their calculation in *Science*, lowering that equivalency, and then Tom Crowther focused public communications on one-third of the total carbon burden in the atmosphere. They were considering what trees could draw down in relation to the excess in the atmosphere from human activity and the fact that trees take up only a fraction of emissions—the airborne fraction, 45 percent. Oceans, for example, play another role. Jean-François promoted an even more conservative estimate of one-sixth in a Ted-Ed video titled "What If There Were 1 Trillion More Trees?" He was using the study's low-end estimate of the amount of carbon that the trees could store, instead of the 200-gigatonne midpoint.**[5]

However, the line in the published study that hit the news and sparked rebuttals across the world asserted that "ecosystem restoration [is] the most effective solution at our disposal to mitigate climate change."

Unlike many other academic research groups in the world, Tom's had full-time staff to help with communications for the many studies the scientists were consistently publishing in top journals. When we later met in Zurich, Communications Director Sam Suarez said, "Having a dedicated communications team is something that scientists don't tend to do." She explained that doing so for pressing ecology-meets-climate-change research was and is important, given the urgency for action and the tendency for helpful research to stay relatively confined to academia. As someone who is also very dedicated to communicating science and problem-solving research, I saw Tom's strategy as bold but admirable and necessary for broadening the reach. I also acknowledged it came with some risks, and it seemed slightly at odds with the cultural norms of the Swiss, who tend to be more private.

The results were twisted and turned into evidence that supposedly alleviated the need to reduce fossil fuel emissions with headlines like

* A gigatonnne is one billion tonnes.

** The correction also explained "the considerable uncertainty range" of 133.2 to 276.2 gigatonnes of carbon "that was missing" from the original publication, given its emphasis on the 200-gigaton estimate (about 735 gigatonnes of carbon dioxide).

"Could 1 Trillion Trees Stop Climate Change?"; "Planting a Trillion Trees Could Be the 'Most Effective Solution' to Climate Change, Study Says"; "Tree Planting 'Has Mind-Blowing Potential' to Tackle Climate Crisis." It was like a child's game of telephone where someone starts with one message and passes it on; by the time it gets to the last kiddo, the original is totally distorted.[6]

Jean-François told me, "In the beginning, we were super excited because when you do a scientific work, in general, it's published, and nobody hears about it, and, okay, onward with business as usual. So, you are tempted to push as much as you can so that just a few people hear about it. None of us could have foreseen the succession of what unfolded."

A group of scientists commenting in the *New York Times* warned that focusing on trees as the big fix was a "dangerous diversion." Planting trees, they said, could contribute to the long-term climate solution, but it wouldn't alleviate the pressing need to reduce fossil fuel emissions.[7]

As the CEO of one NGO told me, "The love affair followed by the breakup was more about the media and how people process information than the science." An exposé titled "Why Planting Trees Won't Save Us" in *Rolling Stone* called the idea that we're going to solve the climate crisis by planting a trillion trees "a particular kind of lunacy." I remember reading that article and thinking, "Whoa, not the typical outlet for in-depth science, climate, and environment coverage; clearly the subsequent tree-planting mania has reached far!" (As a teenager I'd always turned to that magazine for the latest updates on bands like Pearl Jam and R.E.M.)[8]

One scientist I interviewed called the whole mess a "megastorm." She felt frustrated by the amount of energy that went toward trying to address the limitations of the study and the misleading messages: "All those scientists could have been working on other things! So much brainpower that could have been applied elsewhere went to critiques, comments, and clarifications." I also heard the term "shitstorm" used to describe the frenzy. Some critics noted that certain regions identified by Jean-François and his colleagues for restoration had not been previously forested. They claimed calculations in these areas overestimated the potential benefits, given that these lands were considered to contain no carbon.[9]

Other experts were concerned about carbon-focused tree planting in savannahs, grasslands, and shrublands. *Afforestation*, planting trees in areas that haven't recently had tree cover, as opposed to *reforestation*, replanting, could threaten the existing biological diversity in these habitats as well as people who depend on grasslands, for example, to provide livestock forage, game habitat, and groundwater or surface water recharge. (The line between afforestation and reforestation is also arbitrary; it's typically fifty years. Put a forest in a place where trees haven't been present for fifty-one years, and that endeavor could be perceived as controversial afforestation, but pursue reforestation in a place where there were trees more recently, and that's just putting them back. *Forestation* encompasses both approaches.) The article's discussion section had cautioned that restoration efforts "must not lead to the loss of existing natural ecosystems, such as native grasslands," but as Jean-François recounted, "People focus on other things. I tried, but that was missed."[10]

By definition (the Oxford variety), *to restore* means to bring back, to reinstate; to return something to a former condition, place, or position. Ecological restoration is more than planting; it's a process of assisting recovery of an ecosystem that has been degraded, damaged, or destroyed.[11]

"Restoration goes way beyond planting trees," explained Bethanie Walder, executive director of the Society for Ecological Restoration, when we spoke about Bastin's study and the excitement around planting. The organization's mission is "to advance the science, practice, and policy of ecological restoration to sustain biodiversity, improve resilience in a changing climate, and reestablish an ecologically healthy relationship between nature and culture." She suggested that even the Bonn Challenge has promoted planting in ways that can overlook what is required to revive complex forest ecosystems.

"There have been a lot of eucalypt plantations planted in places where eucalypts are nonnative species," Bethanie said. "They might meet the Bonn Challenge or other hectare targets. You can take a fully functioning savanna or grassland, and you can afforest it and plant trees on it. But that's not ecological restoration." *Ecology* focuses on the relation of living organisms to each other and to their physical surroundings. You can't just plant any tree and call it good. That's like trying to restore a painting with a very limited color palette.

"A lot of people were using restoration to talk about *afforestation*, but that's something totally different," Jean-François shared with me. He was referring to the misconstrued sense that the authors of the Global Potential study had suggested that people put forests in places where the land was not previously forested.

Susan Cook-Patton, a senior forest restoration scientist at The Nature Conservancy, modestly calls herself a "glorified carbon accountant." She focuses on analyzing carbon data from forest areas around the world to figure out how much more carbon could be sequestered if we restore tree cover through various approaches, such as natural forest regrowth (letting trees grow back on their own), agroforestry (cultivating and conserving trees in and around crops and pastures), or plantations (growing one or a few species intensively). When we spoke, I laughed when she told me that despite the fact that "restoration" is in her title, she tries to avoid using the term as much as possible.

"Restoration is about putting things back to the way they were," she noted. "But do we even know what that looks like or should look like? Are we talking about putting it back before people were around? Or before industrialized society developed? Or are we trying to think about what the right types of systems are for future climate conditions?" Instead, she opts to use the phrase *restoring tree cover* because then it's clear that you're putting trees back where they were historically, but you still might not make the *forest* into what it was previously.

In my review of all the buzz surrounding the Global Potential study, it was easy to see that "planting trees" was the most tangible action that most people could glean from "restoration." Concerned citizens want to *do* something that might help. Funders want to support action. Planting a tree *feels* like doing more than just leaving one to grow. "Planting" was also a simple example to give when the scientists were discussing the results with reporters. Reporters tend to highlight a concise message without a lot of nuances. And it was probably easier for most people to imagine the act of planting to expand tree cover than to imagine what the forests might be like in Jean-François's world without us.

I found other critiques noting environmental constraints that weren't considered in the analysis. In parts of Australia and other arid regions, for example, soil salinity and moisture limitations could prevent tree establishment. Robin Chazdon, a leading restoration ecologist and evolutionary biologist, noted that the study didn't consider negative trade-offs; in some places increasing tree cover can elevate fire risks, decrease water supplies, or cause crop damage by wildlife. Other critiques focused on the overlooked lives of about 2.5 billion people who reside in areas eligible for restoration. Clearly, Jean-François's intention to pose a timely thought experiment—this hypothetical situation of our planet without humans—got lost along the way. He wasn't focused on restoration itself, as in what people *should do* to restore forests, but the public perception of the results suggested otherwise. To me, the negative responses seemed warranted given the mounting concern among experts that people would plant trees anywhere and everywhere in the hopes of solving the climate crisis.

When I asked my former advisor in graduate school, Eric Lambin, about the Global Potential analysis, he said that it was a good study, a nice work, fine if framed properly and its limits acknowledged. That doesn't sound like high praise, but I knew from my own experience with Eric that "good study" is as good as it gets in his encouraging but humble ways. He said, "There are many people out there who want to hear, 'Oh fine, good. So, we keep burning coal and oil, and we keep living as we are. But we'll just plant a few trees!" He also said the authors got a bit "overenthusiastic" about their results.

Despite the backlash—what some called a real soap opera in the scientific community surrounding the misinterpretation and misrepresentation of this one climate fix—the many researchers I interviewed about Bastin's study agreed on one very key point: it brought forest restoration, all that it entails, and what it could offer future life on Earth into the limelight.

One other important point to note, as I discovered, is that there were lots of critiques about *how* Jean-François and his colleagues calculated the amount of carbon these trees could store as well as some criticism of that result. Jean-François admitted to me that their approach could have been more rigorous. Yet their 200-gigatonne estimate, when compared to others out there, was conservative and reasonable. It still is.

A study, coauthored by Susan Cook-Patton, revealed similar estimates in 2022 with greater detail. The results established an "absolute reference point" for prioritizing locations and actions, such as improved management of existing forests, to increase carbon storage on land. It's a useful study if you want to know more about the maximum amount of carbon that could be sequestered in boreal, temperate, and tropical forests or see the variation across the top twenty-five ranked countries in terms of unrealized carbon storage potential. (Interestingly, Jean-François and his colleagues had estimated that over 50 percent of the tree-restoration potential could be found in Russia, the United States, Canada, Australia, Brazil, and China, stressing the responsibility of some of the world's leading economies.) To enable a direct comparison with the results from the Global Potential analysis, the scientists considered tree biomass and soil organic matter to estimate the unrealized potential, similarly excluding grasslands. That estimate revealed an unrealized potential of 225 gigatonnes (the study reports petagrams, but they're equivalent)—a value within 10 percent of the earlier, controversial result. Jean-François and Tom Crowther had taken a lot of hits for their 200-gigatonne estimate, partly because they hadn't fully addressed disturbances like fire that affect carbon storage. The 2022 study carefully recognized its estimates in high-disturbance areas as upper bounds with yet-unanswered questions. Nevertheless, the total estimates were comparable. In her measured perspective of the potential for trees to help combat the climate crisis, Susan told me, "Forests are a powerful ally, when we need as many allies as possible."[12]

Scientists I consulted emphasized that the annual carbon sequestration service provided by forests is difficult to pin down exactly; carbon is simultaneously absorbed by oceans and soils and continuously emitted under innumerable circumstances. Nonetheless, estimates are out there. Adding to the difficulty in synthesizing or comparing these results is the fact that some studies report estimates for carbon dioxide and others for carbon stored, especially when their carbon calculations are based on biomass—the mass of a tree's live or dead matter above ground and estimates of its roots below. Scientists may also report carbon and carbon dioxide amounts in smaller units like megatonnes; 1 gigatonne is 1,000 megatonnes, and 1 megatonne is 1 million tonnes, or metric tons.

A 2020 study published in another esteemed journal, *Proceedings of the National Academy of Sciences*, suggested that forests in the United States remove between 540 and 850 megatonnes of carbon dioxide (about 150 to 230 megatonnes of carbon) from the atmosphere per year. That estimate is roughly equivalent to 14 to 17 percent of the country's annual carbon emissions, depending on the year.* Naturally, more forest means more storage and a greater net drawdown from the atmosphere. That same study estimated a potential to increase carbon storage capacity in the United States by about 20 percent by stocking lands already used for timber with more trees.[13]

Susan Cook-Patton at The Nature Conservancy was the lead author on another study, published the same year, that called for restoring forest cover on about fifty-one million hectares of US land (not only on understocked timberlands), a total area about twice the size of Oregon. The researchers estimated that the restored forests could capture 314 megatonnes of carbon dioxide per year (about 85 megatonnes of carbon), meaning the potential increase in carbon storage from restoring forests is more on the order of 40 to 45 percent in the United States.[14]

Some scientists, like Susan and her colleagues, report potential for carbon sequestration in relation to targets in climate policy, like those supported by the Paris Agreement. That international agreement put forth the long-term goal of holding the increase in average global temperature to "well below 2°C above preindustrial levels" and aiming to limit the increase to 1.5°C. Leaders have stressed the importance of achieving the goal by the end the century, which requires immediate emissions reductions and achievement of negative emissions, or drawing more down from the atmosphere.

---

* For example, the Environmental Protection Agency reported 812 megatonnes of carbon dioxide sequestered by US forests in 2020, comprising 13.6 percent of the total greenhouse gas emissions that same year. ("Land Use, Land-Use Change, and Forestry." In *Inventory of U.S. Greenhouse Gas Emissions and Sinks: 1990–2020*. Washington, DC: Environmental Protection Agency, 2022.)

Strategies for accomplishing those warming limits are informed, in part, by the global carbon budget—the amount of carbon dioxide that can be "spent" (emitted) for a given level of warming. The 314 megatonnes in the study led by Susan Cook-Patton are equivalent to about 15 percent of the United States' 2016 commitment to the Paris Agreement. My friend and colleague Rob Jackson is chair of the Global Carbon Project, which aims to establish a knowledge base of the carbon cycle that supports climate action and helps stop the increase of greenhouse gases in the atmosphere. With scientists contributing from all around the world and other partnering research bodies, the group updates and publishes the global carbon budget each year. When I wrote to him and inquired about how much this unrealized potential from forests could help with budgeting by trees sequestering carbon, he replied, "Frankly, we're so close to the 1.5°C that the budget is no longer helpful. We've used it up." We will pass 1.5°C of warming, but forests still have a role to play for the next target.

In its most recent update on the carbon budget, the Global Carbon Project released estimates of 325 gigatonnes of carbon (equivalent to 325,000 megatonnes of carbon, or about 1,190 gigatonnes of carbon dioxide) and 305 gigatonnes of carbon that could be emitted from January 2024 onward for a 50 percent chance of limited warming to 2°C. Rob called the many published estimates and complicated comparisons a "quagmire that distracts us from the fact that we're hurtling blindly past 1.5°C." The comparison of Bastin's 200 gigatonnes (or the subsequent 225-gigatonne alternative for unrealized potential) to a 325- or 305-gigatonne carbon budget still suggests that trees can make a substantial contribution to the next target. Yet how much time will it take to realize that forest potential?[15]

The truth is forests do have an enormous potential to sequester more carbon, but harnessing that potential is far from certain or straightforward.

By the time I traveled to Zurich with Calder and my husband, Matt, nearly three years after the media flurry surrounding the Global Potential study, so much had happened. It was certainly not all attributable to the study, of course, but the timing and linkages can't be ignored either.

Organizations such as One Tree Planted and the Arbor Day Foundation doubled or tripled their number of trees planted around the world. Corporate interest intensified: Amazon founder Jeff Bezos gave millions of dollars the following year to plant native tree species and, later, $2 billion to restore forests. In January 2021, the United Kingdom committed £3 billion to protect and restore ecosystems abroad over the next five years. Ethiopia set a world record for planting in a day with 350 million trees. Marketing plans for $1-per-tree donations attracted (and continue to attract) individuals inspired by the chance to help. The band Coldplay pledged to plant a tree for every concert ticket sold, and the search engine Ecosia committed to using the revenue from ads for planting. By Earth Month of 2022, Amazon customers with an Alexa-enabled device could say, "Alexa, grow a tree," and donate a dollar to plant one.[16]

The sheer ambition of the tree-planting movement also created some strange bedfellows. Greta Thunberg, the US Department of Defense, and the IUCN got behind efforts to increase tree cover. In 2020, then President Donald Trump announced a commitment to the World Economic Forum's Trillion Trees Campaign—a global mission to conserve, restore, and grow a trillion trees—in his State of the Union address. Not long after, President Joe Biden directed billions of dollars to managing forests, restoring millions of acres, and responding to increasing fires as part of his Build Back Better plan. The chief executive of Chevron said the company preferred to return money to its shareholders rather than use it to invest in solar and wind. I caught this on Twitter: "We rather dividend it back to shareholders and let them plant trees."[17]

Tom Crowther had also become cochair of the advisory board of the UN Decade on Ecosystem Restoration, an esteemed and influential role in what the United Nations called "a global rallying cry to heal our planet" around its launch. When I finally had the chance to meet him, he had a remarkable ability to be fully present with me, no phone distractions or watch checks. On each occasion, as we neared the end of our scheduled meeting, he kindly reminded me that he needed to be prompt for his next one and made sure I knew which direction I was headed, then dashed off running. I laughed to myself, thinking of an article in the *Guardian* that had described him as the "Steve Jobs of Ecology." He came across more like Clark Kent hurrying to help save the planet.[18]

The Global Potential study was just one of over 100 that Tom had coauthored by the time it came out. Many of the other studies focused on soil microbes and biodiversity. Some were also published that same year in top journals. "What I really want people to know," he told me, "is that forest carbon is just one piece of work that we do."

The media had turned from articles about planting trees as the perfect or not-so-perfect climate fix to profiles of Tom himself. He is "the ecologist who wants to map everything," *Nature* reported; "Tom Crowther wants to restore the planet, but first he needs to know how many trees, fungi, worms, and microbes live in it." Some scientists called him a "disrupter," "the Uber of the field," scooping up local data from forests around the planet and using machine learning to discern patterns at an unprecedented level of global detail. "Catchy findings have propelled this young ecologist to fame—and enraged his critics," *Science* reported in another profile. While scientists questioned his lab's media strategies for its research, the communications efforts also helped spark the much-needed attention to natural climate solutions—conservation, restoration, and improved land management actions that increase carbon storage or avoid greenhouse gas emissions in ecosystems across the world. These natural climate solutions can also fall under the broader description of *nature-based solutions*, actions to protect, sustainably manage, and restore ecosystems that simultaneously address societal challenges.[19]

Tom told me that he was once asked in an interview if he considered himself a scientist or an activist.

"Clearly both," he recounted, emphatically, in his Welsh accent. "I'm absolutely an advocate, and my science clearly contributes to that." From his perspective, every scientific advance keeps reinforcing the importance of biodiversity for sustaining all life on the planet. "My science is in line with my philosophy in life, which is I want to protect and preserve nature as much as possible. I want more nature on the planet, and I'm completely religiously obsessed with this idea that nature is good." He added,

"Everything that I've ever seen from the scientific research also backs that up."

Tom couldn't have been more welcoming when I showed up in Zurich to meet him and his colleagues. The lab had rapidly grown to about forty students and researchers, in addition to four full-time staff for communications. It had offices in two locations to accommodate all the researchers. One wall of a conference room looked like a modern scientist's version of a Jackson Pollock painting. The white board was splattered with equations, calculations, sketched graphs, and a long lists of terms like *ecosystem structure* and *species* with lines and arrows between them. There was an enclosed porch beside the offices in one location with a crokinole board for game breaks. In the space across the street, Tom had turned his private office into a smashball playroom. There were paddles, balls, and scuff marks all over the walls. He worked in a shared space down the hall.

Given his busy schedule, Tom and I met a couple times that week to allow for layers of conversation—first at a café, next over a long walk to his "forest bathing spot," an enchanting little overlook in the Zurichberg forest on the edge of the city. There he pointed out two trees, perfectly spaced apart to accommodate a hammock; it was his place of recharge.

As a young child, Tom had been obsessed with nature. Growing up in Wales, he never really encountered reptiles, but he would turn over rock after rock with dreams of finding one. In the summertime, his family often visited their friends in Angers, France; they had a little French garden with a small pool in it. While the other kids played, Tom would sit staring at a stone wall in the garden, waiting and hoping for a lizard.

"I remember overhearing the grandmother of our friends asking my father, 'What do you think is wrong with him?'"

"What do you mean, what's *wrong* with him?" his father had replied. "He's in love with nature. What's wrong with *us*?"

It wasn't until he was about eighteen that he found his first snake. "It was unbelievable," he told me, lighting up with the joy of that memory from Corsica. "I couldn't believe the ecstasy I had."

I asked him how he found his way to ecology. That journey began with an early fascination with reptiles, but Tom had struggled as a young student with dyslexia. At the beginning of his second year at Cardiff University, his ecology professor, Hefin Jones, threatened to kick him out of a class of some 300 students for messing around. Jones took him to a pub afterward and asked him what he was doing with his life and why he was interested in ecology.

"I obviously love it, but I'm not into all these books and crap," was Tom's reply, as he recounted the story to me. He was earnest and spoke rapidly, as if the words were always trying to keep up with his brain and enthusiasm.

"Find the bits you enjoy and do those," Jones had advised.

"I don't know why, but he completely convinced me to try," Tom admitted. He wanted to study snakes because of his love for reptiles.

"But Hefin was like, 'Just be aware that if you study snakes, you'll have to catch loads and loads of snakes and squeeze poo out of them and take DNA and take clippings and cut them and do things to them. And that might not be the most enjoyable experience for you.' I thought, 'Good call.' So, I studied fungi. I don't mind so much if I have to cut them. But it was sort of along the lines of, 'Don't study the thing you love because you'll turn it into numbers and break some of the magic.'" He smiled, seemingly pleased with the outcome. "So, I study ecosystems, and inherently if I can work on ecosystems and help protect ecosystems then I help snakes and reptiles too. They make up the magic; they're part of the infinite network of life."

His story reminded me of one of my favorite quotes; it is widely attributed to Senegalese forestry engineer Baba Dioum. "In the end, we will conserve only what we love, we will love only what we understand, and we will understand only what we are taught."* Yet Tom's story also made me think of Calder and his innate affinity for trees, his desire to know them, explore them, and take care of them. I thought of how he had

---

* Baba Dioum's quote is commonly sourced to a paper or speech ostensibly presented in New Delhi at the meeting of the General Assembly of the International Union for Conservation of Nature and Natural Resources, often with the given date of 1968. However, the New Delhi General Assembly took place in 1969. A library and publications manager at the IUCN told me they have been unable to locate an original source for the quote and that a note in the library record states that a volume of speeches and addresses from the assembly was ultimately never published. Nevertheless, it is a great quote.

once stood beside a tree on the playground, defending it boldly from the bigger kids who were trying to break off branches. As a toddler, he'd often ask to return to a little grove near our home, where we'd once seen the morning light glistening on leaves in what he called "the sparkle forest." Does all that come from nature? Or nurture? I wondered if he would lose those instincts and what might happen if everyone rekindled theirs now.

"I don't think climate change is a product of people being intentionally bad," Tom told me. "It's a product of growth. If a swarm of locusts expands and eats all the plants, you don't really blame the locusts. It will affect them in the long run. They'll run out of plants, and their population will collapse, but it's not because they were evil. It's because positive feedback sometimes grows out of control and humans have been incredibly innovative, and those innovations have incrementally increased the chances of exploiting natural resources." Now we need to course correct.

"What I find most devastating about climate change is the inequality of it," he added. "I don't like that rich people in the West get to choose and make decisions that lead to massive inequalities and population reductions in other parts of the world, particularly affecting biodiversity." He noted that he didn't think we'd be where we are today, at the birth of a movement that is increasingly focused on solutions, without all the pressure concerned citizens have put on institutions.

"I don't think you see positive action, like mass positive action, until you see optimism. So, you need the fury to stop the bad systems and the optimism to empower the good ones."

By the time we met, Tom and his colleagues had launched a new online open data platform called Restor (without an *e*) to support members of the growing restoration movement. The Global Potential study, the Crowther Lab's use of big data, and Google Earth Engine had caught the attention of Google's Creative Lab. The two labs—Creative and Crowther—had come together to develop what became Restor. The vision, at the time, was a bit like a restoration version of Google Maps; you open it up, enter a location, and find the details of a restoration initiative, as well as data on local biodiversity, water dynamics, current carbon stored, and potential carbon stored with restoration. One day, Tom clarified that in his ambitious dream for this new system, you might also be able to open it up and trace the product you're buying, like a cup of coffee or article of clothing, to its source and the land management practices there. It would be a way for customers to understand the ecological footprint of the products they purchase. In the year following its launch, data from over 100,000 projects had already been uploaded. It had also evolved to connect local community-driven initiatives with funders, customers, and other collaborators.

"We believe anyone can be a restoration champion," I read on the website.

When I asked about the media surrounding the Global Potential study and the push for planting, he cringed. "I've got a knot in my stomach now just talking about it. I had two versions of sadness. One is 'Oh my God, we've got this incredible opportunity to get a good message out, and it's coming out all wrong.' The other is that I felt very bad for Jean-François. If any scientist in the world sits down with me and challenges me on any of the science that he did, I will defend it to the end of the world. He did great science."

Tom admitted to ruining the rollout with his own naive communication. He wasn't trying to promote planting forests of one species or expanding trees into new areas and ecosystems. "Massive monoculture plantations are the antithesis of nature. Those aren't what we were talking about. It sounds so stupid, and I don't know if anyone really believes me, but I didn't think for a second that when we were saying restoration, the world would only think tree planting!"

Much good came along with the bad. "The conversation exploded out of this," Tom said. "That paper spurred the ecological community to come out and say, 'It's not just about trees; it's about ecological integrity and social integrity." He hesitated for a moment, then asserted, "I'm thinking if I could go back and do it all again and make the same mistakes, I think I should. Even though I look like an idiot, I think I should."

Many years ago, I read a beautiful and powerful book called *The World Without Us* by journalist Alan Weisman. It asked, if we removed humans from the planet, what would happen to all that we've created? In what environmental activist and writer Bill McKibben called "one of the grandest thought experiments of our time," the book meticulously details how infrastructure would collapse; what items would become immortalized as fossils; how floods would erode cities and forests would emerge through crumbling concrete and asphalt. That book forever changed the way I see the world. To this day, whenever I drive into a giant parking lot, I still find myself imagining the cracks that would form in the asphalt and the green sprouts that would inevitably surface if, suddenly, we disappeared. I went to my local library one day to reread the introduction, as I was thinking about Jean-François's intention for the experiment on the tree-carrying capacity of the planet. The prelude reads,

> Look around you, at today's world. Your house, your city. The surrounding land, the pavement underneath, the soil hidden below that. Leave it all in place, but extract the human beings. Wipe us out, and see what's left. How would the rest of nature respond if it were suddenly relieved of the relentless pressures we heap on it and our fellow organisms? . . . For a sense of how the world would go on without us, among other places we must look to the world before us. We're not time travelers, and the fossil record is only fragmentary sampling. But even if that record were complete, the future won't perfectly mirror the past.[20]

I thumbed through the other sections of the book, recalling what places like Manhattan would be like in our absence and what might happen to all the plastics we've produced.

There was one line that I read over and over: "Since we're imagining, why not dream of a way for nature to prosper that doesn't depend on our demise?"

Why not? What's wrong with trying something different? Isn't that what anyone who has latched onto the dreamy potential of restoring forests is doing? I thought back to a conversation I'd had with Patrick Meyfroidt, another brilliant Belgian scientist who studies land use and forest transitions—the intriguing and unexpected shifts from forest loss to forest gain, from deforestation to reforestation. I had asked him about the Global Potential study. Years ago, Patrick had helped Jean-François with remote sensing in one of the classes taught by Eric Lambin, our mutual advisor.

Patrick noted that the study wasn't about addressing economic or political feasibility, public and political will, or potential trade-offs with other services that land, now ripe for restoration, might be providing. What the Earth can fit, ecologically speaking in a thought experiment without people, isn't 100 percent attainable when you put all of us back into the equation.

"To me, it's very far from being the end of the story," he said. "It's just an opening that says, 'Okay, maybe there's this potential.' Now what share of this potential is *real* potential?"

3

FROM LONG AGO

Decades of satellite imagery and inventories of the world's forests have painted portraits of change, the result of human activity in very recent times. Yet the trees and forests captured by any modern technology or on-the-ground efforts are only momentary snapshots of the Earth's deep history. The fossils and petrified pockets of plant matter reveal the stories of life long ago. The past still shapes the present and future too.

If any one period has drawn the most widespread intrigue (apart from the Jurassic, that is, as everyone loves dinosaurs), it is the Carboniferous, which began more than 350 million years ago. Its miry forests, primitive, lizard-like reptiles, amphibians with massive skulls, and dragonflies with meter-wide wing spans capture the imaginations of children and adults alike. Oxygen levels were much higher than they are today, maybe also partially explaining the jumbo size of creepy-crawly insects back then. The curious plant life had a high rate of turnover, meaning vegetative communities changed in waves as new members emerged and others disappeared. Paleobotanists, who study the evolution of plants, continue to debate these details and dynamics. But I look back to that period as proof positive that trees can sequester emitted carbon, given that trees were the source of coal. What happened millions of years ago unexpectedly set the stage for the warming world that we face today.[1]

From the Latin *carbo* ("coal") and *fero* ("to bear" or "to carry") we get *carboniferous*, or "carbon-bearing." The trees of these lost forests and a collision of many factors over millions of years gave this age its greatness. What we think of today as the green above became the coal deep below, to be discovered much later, then prized and pursued. Our machines mined the black gold and put smoke in the sky, revolutionizing the world and altering the atmosphere.

When I asked Kevin Boyce, a renowned paleobotanist at Stanford, for the best description he's come across of the forests from the Carboniferous, he laughed and said, "Dr. Seuss. You know, *The Lorax*." I chuckled too, remembering the lollipop trees from the classic Dr. Seuss book from the 1970s that chronicles the plight of the environment. In the story, the Lorax character "speaks for the trees" and confronts the Once-ler, who clears all the Truffula trees to make Thneeds, these things that no one needs.

"Except, you couldn't make thneeds out of those," Kevin added, smiling.

But we did, kind of. We burned those trees—billions of them, maybe more—after they became coal.*

What a forest was then is not what a forest is anywhere now. Green stalks stood like colossal asparagus spears in a swampy landscape; they could reach over 100 feet in height and 6 feet in diameter at the base. These were the arborescent lycopsids, members of a group of the oldest vascular plants.

Those enormous trees have only minute relatives living today. I mainly know them as members of the *Lycopodium* genus, but there are others too. I spent a few summers studying plant communities in Alaska for my PhD about a decade ago; those clubmoss relatives with creeping stems were the tiniest species that we recorded. In the Carboniferous, the lycopsids grew tall, making for the sky and spending much of their lives as verdant poles with thick, scaly bark, like the skin of a flightless dinosaur.

These Dr. Seuss–like forests were replete with intriguing characteristics that have been revealed over time through well-preserved specimens found fossilized underground. The complex root system of a "scale tree"

* This is not to say that there haven't been enormous benefits to people from the Industrial Revolution, but the thneeds get at the troubling inequities that have also been created and the differences between *needs* and *wants* when it comes to consumerism.

had tubular structures woven underground; they could sprawl over an area nearly eighty feet in diameter, binding sediments below and offering stability to the trunk above. Unlike many trees today, these individuals didn't offer much shade. Some of the lycopsids reproduced just once in their lives. Their stubbly crowns unfurled only as they approached their old and dying days. They grew, grew, grew, reproduced, and then died in their grandeur.

Reading paleobotany articles about these trees, trying to wade through descriptions in another language ("lateral plagiotropic branches" and "sporophylls with prominently leafy distal laminae"), made me think of Calder's simple question when we were at Kew Gardens in London. "What do they need to grow, Mama?"

They looked different from the living trees that he and I now know, but they were still lungs for the planet, drawing in carbon dioxide and releasing oxygen. Their basic needs for growth were like those of modern plants: sunlight and water and carbon dioxide to make sugar—their food and energy. Relatively speaking, they had poor plumbing systems. Leaves in modern trees are the critical player in photosynthesis; the innermost part of bark, the phloem, helps transport food made in the leaves to all the other parts of the plant. But the lycopsids of the Coal Age likely did that locally with leaves instead concentrated near specific areas of growth. Those extensive root structures may have also scavenged carbon dioxide from the soil, whereas trees today act as the lungs from the canopy.[2]

Between about 330 million and 260 million years ago, the highest rates of global organic carbon burial over the past half billion years occurred. That was due, in large part, to the accumulation and burial of peat—the rich, black and dark brown spongy material formed by partially decomposed organic matter in low-lying basins.* For about nine million

* There were also other plants that made up the forests of the Carboniferous period. There were early tree ferns, related to some modern ferns with giant fronds that one might find in the tropics today. There were cordaitales, a sister group to our modern-day conifers. They had elongated, strap-like leaves and lived from the edges of the ocean up to high altitude, as well as in habitats ranging from swamps to seasonally dry areas. There were some of the earliest conifers later too. (DiMichele, William A. "Wetland-Dryland Vegetational Dynamics in the Pennsylvanian Ice Age Tropics." *International Journal of Plant Sciences* 175, no. 2 [February 2014]: 123–64. https://doi.org/10.1086/675235; Cleal, Christopher J., and Barry A. Thomas. "Palaeozoic Tropical Rainforests and Their Effect on Global Climates: Is the Past the Key to the Present?" *Geobiology* 3, no. 1 [2005]: 13–31. https://doi.org/10.1111/j.1472-4669.2005.00043.x.)

years, the bark of the lycopsids was the single greatest contributor to what became coal.[3]

Even after all the reading I've done about the forests of long ago, I still have to remind myself that "then" is on a timescale very difficult for most human minds to fathom (mine included). Just as "now-ish" for some paleontologists can mean the last fifty million years, "then" occurred across millions of years. The trees didn't all fall down for burial in one dramatic event and then (ta da!) there was coal. Instead ideal conditions came together repeatedly across time. Wet tropics fostered productive forests. There were rain and swamps and cool conditions. Glacial periods came and went, intermittently flooding and drying out the landscape. And the bark of those trees was like concrete, more resistant to decay than anything else. Bill DiMichele, a paleobotanist at the Smithsonian Institution's National Museum of Natural History who has spent his career studying the Carboniferous, noted that the lycopsid bark was "the plastic of its day," accumulating even when everything else was decaying away. During the wettest of times, what got buried below faced a waterlogged world and prime conditions for preservation.

The configuration of the continents was also key. Lands that we know as separate today were contiguous as one massive terrestrial surface—Pangea, the megacontinent—creating opportunity for burial.

"It was a bunch of continents banging together, which makes mountains," Kevin told me. We were meeting in his office on the Stanford campus. In the adjacent meeting room was a shelf full of coal balls, black lumps of petrified plant matter found in coal beds—each sample discovered in serendipity, each offering clues to the past.

Whenever mountains uplift, basins form. So, there was this perfect intersection of climate and other factors that enabled tropical organic matter to be buried and preserved. The basins would fill up just as fast as they were subsiding. All of this happened very, very slowly in terms of human time (as opposed to geologic time).[4]

"It's like you're accumulating centimeters per century," Kevin explained. "Basically, it would never look like anything at one point in time; there's no hole to fill. It's just the basin bottom dropping out as it's filling at the top, so that stuff is accumulating rather than being eroded away.

There are sedimentary basins in the world with ten kilometers of sediment in them, but that doesn't mean there was ever a hole ten kilometers deep."

As the basins formed, there was always something to fill them. The carbon-rich trees played that role, becoming buried treasure for people millions of years later. The use of that buried treasure would also become a leading contributor to the unexpected calamity—the warming world. Coal is the dirtiest of the fossil fuels on many levels; its consumption also has detrimental effects on human health. Between 1850 and 2020, almost half of global CO_2 emissions came from coal, more than any other single source.[5]

Now what? People are turning to trees again but, this time, to their thriving verdant form instead of the coal formations that they helped create. They want more trees to absorb carbon dioxide from the atmosphere. We've come full circle in a relationship between people and trees—from where using what was buried became a source of emissions to where the

Carboniferous coal swamp.

Credit: Hannah Bonner, created with guidance from Bill DiMichele at the Smithsonian Institution. From the book When Bugs Were Big *(National Geographic Kids, 2015).*

already-standing or freshly planted trees could be our saviors, drawing down carbon through a natural solution. People are chasing after a hope-filled possibility that trees might just help get us out of this mess.

The Earth has always been in some sort of disequilibrium with volcanic eruptions, rock weathering, changes in sunlight, and other forces driving changes in the climate system. But human activities are pumping about 100 times more carbon dioxide into the atmosphere than volcanoes do each year; it's the rate and magnitude of the current disequilibrium that puts the lives of people and many other species in grave danger.

"As long as we're in disequilibrium, which we are, because humans are a source of disequilibrium, then growing forests matters," Kevin said.

There is no way for trees to offer the whole solution, but I agreed with him. They have a role to play. Coal and the carbon that it contains accumulated over tens of millions of years. Forest clearing by humans occurred much later in a relatively teeny sliver of time. The burgeoning interest in planting or replanting forests arises in an even narrower window. Those involved, myself included, sense the clock is ticking.

4
SOMETHING NEW UNDER THE SUN

I first heard the term *forest transition* when I was a graduate student. Sometime during my first year, I'd read a paper that described the phenomenon as an "intellectual shorthand for a historical generalization" about changes in forests and their surrounding human societies over time. "The point of inflection in a transition," it explained, "occurs when deforestation disappears and reforestation commences." I don't think that the concept of a reversal fully captured my attention until years later when I was walking through California chaparral with Eric Lambin, my former advisor and a geographer who studies the factors that drive people to clear forests or to plant them again.[1]

It was a hot, dry day in June 2019. Even the manzanita appeared parched. Oak leaves crackled under our feet. By then I was already hooked on the questions "Can trees really save us? And how?" Like the many other citizens and scientists concerned about climate change, I was focused on carbon. I was wondering if the growing interest in natural climate solutions could effectively spark a shift away from forest clearing to more forest recovery.

Eric cautioned me not to see a forest as only a carbon sink or carbon stock. "The biodiversity aspect is really important, and increasingly, it's also linked to the carbon," he said. *Biodiversity* encompasses everything from variations in genetics to all species, including plants, animals, fungi, and even bacteria and viruses. "The carbon sink function of a forest is much greater, and the forest is more resilient to climate change, when there's some biodiversity associated with it."

He'd spent much of his early career studying the causes of forest decline: what factors—like population density and distribution, market growth, agricultural expansion, road development, policies, or even public attitudes and beliefs—lead to deforestation? Given his expertise, people often asked him, "How can that understanding of the causes inform policy design to help reduce deforestation or offer clarity on what levers could slow it?" Eric started thinking a lot about that question, but any insights he could offer were largely theoretical because they assumed that trying to address the causes of deforestation could create more forests. In reality, increasing forest cover or attempting to restore forests might require something different.[2]

"You have to look back twenty years, where it was all doom and gloom over land use, climate change, and biodiversity loss," Eric said. "In a sense, that's still here today, but at that time, that was the dominant discourse. As an environmental scientist, that's what you were doing; implicitly, your job was to chronicle or to document the world that was transforming and try to measure that and inform the public and policymakers of all the consequences."

Decades earlier than my inquiry, he'd also been searching for hopeful stories of recovery. Around the time he published his work on the drivers of deforestation, he met a researcher named Alexander Mather at the University of Aberdeen in Scotland. That was 2002. Mather had studied the historical forest transition that had occurred in Scotland, where he was from, and then other such transitions—from forest loss to forest gain—throughout Europe. Eric became interested in similar situations of increasing forest cover that might be occurring more recently in other countries.

National-level forest statistics revealed such reversals in Costa Rica, Vietnam, Bhutan, and Chile. There were (and now are) others, but those caught Eric's attention first.

"I moved away from the mode where my job was to document what was being destroyed. My job was also to identify what were the leverage points where we could intervene and turn things around," he told me.

Eric focused on reversals in certain countries, but I was wondering more about one at the larger scale. How could people harness as much as possible of that *global* potential?

"Do you think it takes leaders in each place?" I asked him. "Or is each local story its own model?" I was thinking about planting efforts and restoration projects as well as the prospects for halting forest degradation in exchange for better stewardship and care.

"Every effort is contextual," he told me. "There are some generalities, but there is no one-size-fits-all solution. No one ever said, 'Oh, let's go into a forest transition now.' It happened because there were a number of factors that aligned and converged."

I thought that looking back at what Mather had learned from historical forest transitions might offer insight on the path toward the regrowth and rebirth of our world's forests today.

Mather died in 2006. But he left behind a trail of papers on reversals of the past, foreshadowing so much that was yet to come for the growing global movement.[3]

A somewhat hidden figure in the recent history of our world's forests and their comings and goings at the hand of man—that's how I've now come to think of Alexander Mather, or "Sandy," as he was called by those who knew him. I tracked down his former colleagues and spoke with his son. I found obituaries and tributes offered in Scotland around his death and a few articles about him in various papers. I assembled a portrait of a caring and insightful man, a bit dour yet dedicated and unassuming, driven and remarkably productive in his scientific research. The northeast corner of Scotland, where he was born in 1943 and where he stayed, is not famed for a culture of openness, and those who knew him say Sandy was no exception.

"Sandy was a true North Easter in the sense that he had a pawky, dry sense of humor, and never blew his own trumpet nor wasted any words," Jamie Fairbairn, one of his former research assistants, shared with me. "He built up his great knowledge, his reputation, and his status purely through the integrity of his work and staying put in one place. In that sense he fulfilled the Scottish Geddesian tradition of internationalism, looking outwards while having a good grounding and knowledge of his own culture."

Scotland's forest transition wasn't an outlier in the world, but it was what Sandy saw first—a green reversal and a legacy of change left on the landscape of his homeland. After the retreat of the ice some 11,000 years ago at the end of our Earth's last ice age, boreal species became the early verdant colonizers of the freshly exposed terrain in Scotland. More temperate species came next—the beautiful birch, followed by hazel, pine, and oak trees too. Wildlife communities, rich soils, and the structure and shape of the long-lived forests took time to develop. Scotland became mostly forested, as woodlands reached far to Shetland and the Western Isles. But by around 1600, only about 4 or 5 percent of land still retained those ancient woodlands.[4]

"There is no doubt that the forest . . . [had] shrunk to a negligible extent by the 17th century and remained so for the next 200 years," Mather wrote in 2004. By the end of the 1980s, the forested area in Scotland had risen back to some 14 percent. He'd spent years studying why.[5]

I found many other examples of countries, particularly in Europe, that had undergone historical forest transitions. In the Netherlands, forest area grew by almost one-third between the end of World War II and the mid-1980s. Forests occupied nearly 20 percent of Switzerland in 1863; by 1983, that estimate had risen to 32 percent. In Denmark, forest area had almost tripled between the beginning of the nineteenth century and the end of the twentieth. The cutting of forests in seventeenth-century Japan inevitably triggered increased flooding and soil erosion, as well as shortages of timber and fuel, but policies then encouraged planting. By the end of the twentieth century, nearly two-thirds of the country's land had tree cover. China and the United States provided other examples. New England experienced a forest increase during the 1800s. I was more familiar with the legacy of President Franklin D. Roosevelt's "tree army": workers from the Civilian Conservation Corp (CCC) and the Works Progress Administration had planted more than 220 million trees between 1933 and 1942 on the Great Plains to slow the wind erosion that caused the Dust Bowl; the CCC had employed millions of men and planted an estimated three billion trees across the United States. Alexander Mather had begun by examining what occurred in Scotland. Then he looked beyond, searching for broader patterns from the past.[6]

Unlike Matt Hansen and the many other researchers working with satellite data, Sandy Mather wasn't trying to map the physical changes in forest cover across the planet. He searched instead for changes in policies, economic conditions, and even migration—factors that could contribute to a change in land use. Mather was after the root causes of the new patterns on the landscape. He wanted to decipher how changes in human relationships with the land and forests, policies, and other factors have triggered reversals.

What was driving the increases in forest cover?

Mather described the loss of Scotland's forests as relatively early and notably severe. The causes were diverse: clearing of the land for agriculture and livestock and overexploitation for timber. They were common causes of deforestation in Europe and other parts of the world, but Scotland's cool and damp climate accentuated the downward trend. Once cleared, trees had a hard time regenerating in places, and growth was slow going. Seedlings stalled in wet conditions. But the point at which the downward trend shifted to an upward trend was also easily identified. With timber resources declining, the United Kingdom had turned toward imports, with more than half coming from Russia. Not only did World War I trigger a fear of shortage, but there was also concern over depletion of forests in other supplying countries. Planting more trees could increase the domestic supply and decrease reliance on other nations. On September 1, 1919, the Forest Act came into force, and with it a Forestry Commission responsible for looking after the woods across England, Scotland, Ireland, and Wales. Then the expansion began.[7]

As Mather wrote, "Little attention was devoted to the conservation of the limited areas of natural woodlands (at least until the latter part of the century): they were usually regarded as 'scrub' that would not produce the type of timber now urgently required." Scotland wasn't regrowing what once was; Scotland was creating anew—plantations that would yield timber.[8]

"If we simply lump everything together that has trees, then a lot of places look like they're going through a forest transition when it's really an expansion of productive landscapes like plantations. I don't know if you want to call that a forest," mused my friend Robert Heilmayr, an ecological economist and assistant professor at the University of California, Santa Barbara. (Notably, it wasn't until 2022 that a team of scientists published a first wall-to-wall global map of forest management, distinguishing naturally regenerating forests, for example, from plantations and agroforestry, the integration of trees and shrubs into farming systems. That map certainly wasn't perfect either.)[9]

When Mather turned his attention from Scotland to other developed countries that had increased their forest cover, he and his colleagues saw distinct pathways to these historical forest transitions. In one driven primarily by economic development, industrialization and agricultural intensification led to people abandoning less productive farmlands. Often on hillsides surrounding productive valleys, the forsaken farms could revert to forests. Many did. In another pathway, what Mather called the "crisis-response model," scientists and policymakers who perceived immediate threats from deforestation responded with conservation and reforestation measures or policies, like what had occurred in Scotland. In Switzerland, fears of floods and wood shortages had inspired efforts to expand forests. In Denmark, a growing population combined with increasing agricultural land use and demand for wood had triggered its forest-resource crisis. Other researchers later expanded on this model by identifying how fears of dependency on foreign resources, such as timber, during World Wars I and II also motivated new planting programs.[10]

A legacy of unsustainable land use was always underlying the reversals in forest cover that Mather had studied. Ensuing problems from great loss became a stimulus for action. They helped chart the new course toward a forest transition. The many scientists and colleagues that I interviewed about Mather left me feeling that he spent his life working on a topic that was, back then, fairly academic and theoretical. Then it became one of the most pressing of our time.

From Mather's work, I could see that the real potential of forests is about not just where the trees could be and what they could offer but the ability of people to chart a different course. It is about people figuring out how to keep forests thriving across time through localized efforts at a global scale.

My husband, Matt, comes from a family of writers—generations of writers, in fact. I fell for him long before I ever saw the library in the home that he had inherited from his grandparents. In those early days of getting to know each other before we had Calder, I admit that I was drawn to the idea of writing in a room lined with over 5,000 books in different languages. There were a few shelves devoted to the writers from his family. We began reading to Calder as soon as we brought him home from the hospital. We very briefly entertained the simple black-and-white picture books recommended for newborns and then moved to full narratives. Matt would recite Hemingway from memory while pacing with him late at night until Calder fell asleep. I began reading to him children's books about scientists and bugs, about climbing mountains and planting trees.

It was a snowy winter morning, and Calder and I were snuggling in a cozy tent on the floor of his bedroom. The tent was white and black and shaped like a rocket ship with classic NASA lettering on the outside. We barely fit inside together.

"It's more of a pod than a full rocket ship," he often clarified for me. "Just one astronaut at a time!" When we read books and told stories in there or made occasional trips to the moon, my legs extended out the fabric flap door.

On that day, we were reading *The Lorax* for the first time. Calder had picked it out at the library, and I'd hesitated before agreeing to take it home. I remembered reading it as a child and feeling sad for the birds and the Truffula trees, though I doubt I made any connections to the real world at that time and most certainly didn't know the resemblance to any trees from millions of years ago. I was afraid too many questions about forest loss might come from my inquisitive three-year-old.

At the end of the book, the Once-ler tosses the last Truffula seed to a little boy who comes upon the cleared landscape. The Once-ler explains to the boy that he is now in charge of the one remaining Truffula seed and that Truffula trees are what everyone needs. He tells him to plant a new Truffula, care for it, and give it water and fresh air. "Grow a forest. Protect it from axes that hack. Then the Lorax and all of his friends may come back."

I was accustomed to Calder quickly moving on to the next book after finishing one. But he lay quietly beside me, thinking for a few minutes.

He nuzzled closer and asked, "If you plant a tree, will it make others? Will everything be how it was again?"

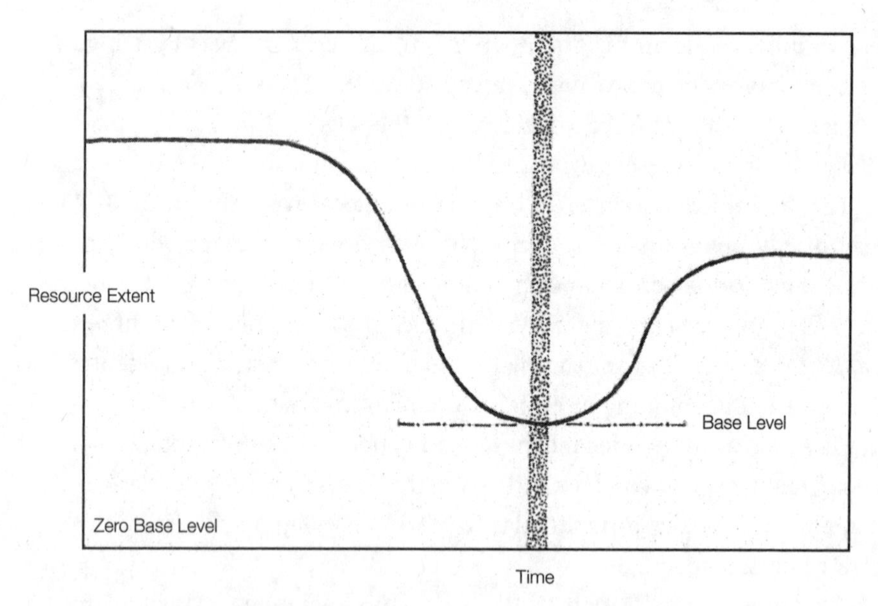

Stippled band indicates forest transition.

Credit: Lorenzo José Rosenzweig, adapted from figure presented in Mather, Alexander S. "The Forest Transition." Area *(1992): 367–79, after Whitaker, J. R. "World View of Destruction and Conservation of Natural Resources."* Annals of the Association of American Geographers *30, no. 3 (1940): 143–62.*

In one of the many articles that I read from Mather's prolific career, I came across a figure that immediately caught my attention. It is a graph, with resources such as forests on the *y* axis and time on the *x*, then a simple black line drawn like a profile of two rolling hills with a valley between them. The second hill is smaller than the first. Sliced straight through the valley is a stippled vertical band that divides what once was and what now is.[11]

Mather called it a "depletion-melioration model." The language struck me, a trained scientist, as intense academic jargon, but the idea resonated.

We use what we can until it's blatantly obvious that we need to make a worsening situation better. Now. Urgently now.

In the time represented by that band, something dramatically changes in our human behavior, in our use and values, perhaps even in our relationship with nature. His ideas of how we deplete and then attempt to fix echoed back to other writings about environmental destruction from early in the twentieth century.[12]

Native Americans, Alaska Natives, and other Indigenous peoples have a long tradition of living sustainably with the natural world by preserving natural resources, avoiding overuse, and generally respecting the interdependence of living things. As the graph and discussion in Mather's article suggests, every country, every place, every community has a resource extent that can vary in abundance, or "extent," over time. In a more colonialist model, any unsustainable relationship with nature—or what most people commonly call "natural resource use"—drives depletion. Yet, perhaps in that low of the resource extent, there is a turning point where people change their actions. What was lost is still needed, but it's not all gone forever. Recovery becomes essential.

I thought of Tom Crowther and the locusts—a population booming, eating all the plants, and then collapsing. Plants regrow. Population crawls back up. The cycle begins again. For humans, destruction might be a precursor for conservation and restoration. Perhaps humankind can approach the brink and then draw back. When we'd met, he'd told me, "I don't think we're all doomed. Nature will recover." But the essential question he presented was, "Will the planet recover to a state that supports the life that we see *now*? Or will it be a totally different state?"

For quite some time, I stared at the line forming a trough between two hills and that stippled vertical band cutting through its nadir. Somehow such a simple sketch, generated from ideas put forth over a century ago, captured so many of the questions running through my head about the world's forests.

Mather was focused on forest transitions that occurred at national scales. Is a global one happening now? Could one occur in the future? And might it unfold at the rate and scale needed to harness the potential of trees, not only to sequester carbon but to protect more populations from collapse, including our own? He had described what occurred in that stippled band in Scotland and in other countries in the past, but I wondered if the movement today was unfolding as something different.

Had that second, smaller hill ever fully restored what was lost from the first one? Did the right side beyond that vertical band equal the left side? Could it ever equate in the future?

Between visits with Tom Crowther in Zurich, I met with Rachael Garrett, a former lab mate of mine at Stanford and a leading environmental social scientist at ETH-Zurich. (She accepted a new appointment at Cambridge soon thereafter.) Rachael has spent lots of time in the Brazilian Amazon over her years of working with big agricultural corporations, governments, and communities to try to reduce deforestation and improve the lives of the people residing there. I've always appreciated how she can pull off a sharp suit when necessary but otherwise seems to prefer jeans and a casual T-shirt. She is well-spoken, unpretentious, and dedicated to curtailing deforestation.

In high school, a trip to Costa Rica had showed Rachael how local people could cut down primary forest, only to create agricultural systems that didn't end up supporting them longer term. They had lost their forest, and they didn't gain what they were expecting. By graduate school, she was organizing special seminars in environmental ethics with Nobel Prize winners and engaging students and faculty in important conversations about the unexpected consequences of land use. She was concerned about the inequities that come from environmental degradation. How people value nature—and forests in particular—was as much about ethics as it was about land use and economics.

We sat outside at a corner café, talking about the state and fate of the world's forests as cars whizzed by. The cuffs of her shirt were rolled just high enough to expose a tattoo of tree foliage extending toward a yin-yang sign on one forearm.

We had exchanged a few emails in advance to coordinate our timing around my meetings with Tom. "So, do you buy what Tom is selling?" she had written in her direct style. I had laughed at the wording, as I hadn't really thought of him as someone *selling* a product. But he was, in fact, promoting ideas about what our world's forests could be and how, and what that might mean for future life. Those ideas, backed by data, did include planting (among other approaches, like natural forest regeneration). And let's just say there were certainly a lot of governments and corporations and political leaders *buying* those ideas, along with many other similar ones out there, as they shelled out money and announced reforestation commitments.

I didn't tell her this then or when we met in Zurich, but I had found his vision for the online Restor platform that supports and promotes restorative land use across the world to be inspiring. I saw the potential benefits of being able to open an app inside a store and consider the land-use practices of a coffee company before buying its beans, for example. It's easy to forget that even Google Maps was clunky to start, given its scale of usage and ease of function today. *Maybe with time this could be something impactful?* I also knew that linking one product to one piece of land through a complicated global supply chain wasn't the only challenge; another big one would be tracing a product to what happens on the land and in a community or many communities. Even if that could work for everything we "resource," convincing millions of people to make the practice and decision making a part of their daily lives would be another giant, and perhaps insurmountable, hurdle.

Rachael's strategy, in contrast, is to work with the big agricultural corporations themselves and try to make unsustainable land-use practices slightly more sustainable in ways that affect millions of hectares of land and millions of farmers. She calls it an intermediary approach because making our current commodity system "a bit more sustainable isn't actually sustainable"; we need a radical transformation to a system that is not structured around resource extraction.

"In Brazil, so much of the economic growth is based around commodity agriculture, and that is something that is in tension with the standing forest," Rachael said. "You need to develop an economy that benefits from a standing forest, or an economy based around bio products, nontimber products, around carbon, around biodiversity." She mentioned possibilities for cosmetics, vegetable oils, and even seed collection by local Indigenous people as economic opportunities for nontimber products.

Tom had called this "economically empowered conservation," but he saw it coming organically from concerned citizens across the world. Rachael sees it as a more intentional policy process to change the social and economic systems that guide people's actions and choices. We probably need both—grassroots movements and policy changes.

I took a sip of my coffee. "You asked me, 'Do I buy what he's selling?' But what does that mean to you?" I deflected. "There's lots of money going toward restoration efforts now, but do you think that should be going only toward avoiding deforestation? Is restoration a worthwhile endeavor too?"

"It is a worthwhile endeavor, but any definition of restoration has to have justice in it."

I wanted us to pause for a minute, as I tried to sort out what that encompassed. Maybe she could see my wheels turning and the need for more clarity on how such a definition could play out in practice. I knew the relevant concerns surrounding the misleading message of "plant trees to solve climate change." Someone in one place can't just decide trees *should be* in another place. So much land is in use that either protecting or restoring forests needs to be economically viable for the local people that use, own, or rely upon that land one way or another.

"So, if you're talking about a goal for justice-centered restoration," she said, "which means bringing ecological functioning back to a landscape while supporting the livelihoods of the people who live there and including them democratically in the system, then yes, I support that goal and money going toward that goal—as long as the people there are involved in figuring out how to distribute it." Rachael continued,

> I think a big problem with all of this is how are we defining our goals. On the surface, they are ambitious and exciting: plant a trillion trees; restore 150 million hectares; end deforestation in our supply chains. But how do you do it? You're not saying how to do it. You're not saying the rules that you're going to follow, and you're also not saying how you're going to promote well-being in society. So, if you're ending deforestation by excluding people from lands, then what are those people going to do? Do they have an alternative livelihood or are they going to keep doing what they're doing? Are the trees going to keep standing, if the underlying pressures there continue to support clearing them?*

* The year after we met in Zurich, Rachael and her colleagues published an article showing that approximately 1.4 billion people, disproportionally belonging to groups that score low on measures of human development (i.e., an index of income, life expectancy, and expected or average schooling), live in

I thought of Alexander Mather as she spoke and of how he'd seen changes in policies to support more forests in Scotland.

We use what we can until it's blatantly obvious that we need to make a worsening situation better.

Except this time the worsening situation is not in just one place, in one country or community; it is at the global scale, as temperatures rise and other consequences ensue. In the same vein as Tom Crowther and his infectious optimism, I'd like to believe that perhaps we're at a moment for a global shift where something new and great could happen, and is happening, in that stippled band. Despite persistent naysayers, more people are becoming concerned about climate change.* Whether we're drawn to forests for their timber or their carbon, the habitat they offer, the air they create, or the shade or even feelings of comfort and calm they impart, maybe we can, in fact, care for them differently as we find our collective way forward.

"As long as we continue to lose primary forest," Rachael asserted, "I don't see how anyone can claim there's a global forest transition. I don't know if it's useful to focus on global anything because what matters is the distribution of where we're losing forests, what we are losing and what we are gaining, and how long it will take for a recovered area to actually in any way resemble what we lost."

Mather had been rooted in place and spent much of his life around Aberdeen. Yet eventually his work on understanding the causes of forest transitions caught the attention of a growing body of researchers who called themselves land-use and land-cover change (LUCC) scientists. From 1999

areas that researchers have identified as high priority for restoration. The authors argue, among other key points, that restoration should always be guided by local knowledge and prioritize the perspectives of the most vulnerable people. As their case studies suggest, restoration can lead to improved quality of life for local communities alongside ecological goals when livelihoods and equity are considered throughout the process. (Löfqvist, Sara, Fritz Kleinschroth, Adia Bey, Ariane de Bremond, Ruth DeFries, Jinwei Dong, Forrest Fleischman, et al. "How Social Considerations Improve the Equity and Effectiveness of Ecosystem Restoration." *BioScience* 73, no. 2 [February 1, 2023]: 134–48. https://doi.org/10.1093/biosci/biac099.)

* For example, the Yale Program on Climate Change Communication releases annual "Climate Opinion Maps" with estimates of US climate change beliefs, risk perceptions, and policy preferences. The 2021 data revealed an estimated 72 percent of adults believe global warming is happening, and 65 percent are worried about it.

to 2005, before I studied with him at Stanford, Eric Lambin served as chair of the LUCC; the effort to study global land use was born from a recognition that people were changing the surface of the Earth all too quickly with persistent degradation and that the relevant research tended to be isolated. Gaining a better understanding of any patterns across the world required synthesizing and integrating information in an unprecedented way. Eric recalls that bringing together different perspectives and expertise was central to the endeavor. There were NASA-funded scientists with "wall-to-wall mapping of the Amazon and ten years of data" coming together with Mather, who was shy and humble and doing more "old-fashioned work." They worked together to develop powerful ideas about what was motivating people to change from a relationship centered on forest destruction to one that prioritizes regrowth.

Certainly analyses have shown an increase in tree cover at the global scale. Xiaopeng Song and his colleagues offered one of them. But with a closer look, there was more afforestation, reforestation, and natural forest expansion in temperate, subtropical, and arctic regions than in the tropics, where there's still massive deforestation. A boreal forest is not the same as a tropical one; people can't just swap one for the other and call the exchange equal.[13]

"The statistics are about *tree cover.* They're not about *forest,*" Eric told me once on one of our many walks. I appreciated the precision in his distinction, his willingness to see a forest as more than the trees, to draw that line in more than just philosophy but in science. Estimates might include urban trees and plantations, large monocultures in Canada and Scandinavia.

"I would not talk about a global forest transition," he cautioned. "It would lump together such different ecosystems that I think it has no meaning what-so-ever."

In recent decades, there's been a major drop in global poverty estimates, but China has been at the core of that shift. The country's population is so big that pulling nearly 800 million above the International Poverty Line can swing the global figures. But this success has not been uniform worldwide; rates of poverty in sub-Saharan Africa remain stubbornly high. What's changing in China can mask sustained problems in Nigeria or Zambia when data are consolidated into a global statistic. That's also true of forests or any other critical issue. The vastness of the data can obscure the nuances between one spot on the planet and another.[14]

"Forest gain is just a total hodgepodge of all kinds of things," one ecologist told me when we spoke about forest transitions. "The drivers of plantations are going to be different than the drivers of agroforestry, or natural regeneration, or active restoration and planting." Teasing apart the local drivers of a global pattern requires "disaggregating forest gain."

National policies, for example, can trigger increases in tree cover or forest cover.[15] But farmers and forest landowners and community members on the ground also need to make decisions about what that burgeoning verdant landscape will be, how the trees will get there and stay there with support and care. Otherwise, the risk is that local people will lose out for the benefit of others far away. Ecological economist Robert Heilmayr told me,

> Part of the reason "trillion trees" has taken off is because it can be everything to everyone. The environmentalists see what they want, and then you get Republicans that can see it as a big push for industry. Then once you get into the details of defining what are the values you want any forest to support, they are often in tension. You can't necessarily achieve all of the carbon, biodiversity, economic outcomes, and local community development benefits at once; it's hard to get them all from the same forest you're planting or restoring. There are different forests that provide those values. People are often talking in these very loose terms that prevent them from grappling with that difficulty.

When Calder was just fourteen months old, I took him to see some of the world's largest remaining trees. I knew he wouldn't remember the experience, but I did it for me too. I wanted to witness his response, to share in the feeling of walking beneath ancient giants together. We went to Three Rivers, California, a tiny town on the edge of Sequoia National Park. I had been asked to give a talk about climate change and forest management in the community and to meet with researchers. If there is a moment when

I was fully present in a forest—coexisting in reverence and thinking of nothing in this troubled, warming world—it is when I stood under the sequoias with my son and my mother, who had accompanied us as well. Three generations of humanity, a speck in time against the millennia of massive trees.

"Whoaaaa," he exclaimed, pointing upward in elation, then falling silent to wonder beside the largest-known living single-stem tree on the planet, General Sherman. During the rampant fires of 2021, it was the same tree that fire fighters would later wrap in aluminum for protection from the flames. I stood him on the ground beside another trunk; he touched its bark and craned his neck so far that I thought he would topple over. *Awe is instinctual,* I thought, *and so is survival.*[16]

Even amidst all the past destruction, the clearing of forests, and the reshaping of the Earth's surface, I'd like to believe that people have never truly lost that sense of awe, that wonder about and admiration for big trees. Perhaps people have always been drawn to them just for what they can tangibly offer and what we know at the time they might provide. Warmth, through fire. Transport, by canoe. Storage, with baskets made from bark. Eventually, the roofs over our heads and walls for shelter. The air we breathe. A diversity of life. And yes, the ability to suck up carbon, which most people might never have known we would need so desperately, had we not gone so far in reshaping our atmosphere.

I don't think that second hill in the "depletion-melioration model" will ever be what the first was exactly. But that doesn't mean creating it, or attempting to do so, isn't a worthwhile endeavor. The trees we work to keep standing or let grow or make grow may help make the planet become a bit more habitable in the future; they could still add up to something new, something better than where we're headed without action.

The scientists I interviewed who were familiar with Mather's work referred to "a Mather-like transition" when characterizing what had happened in some historical reversals. One main driver was the migration of people; forests regrew when people left upland regions and moved to cities. Yet this green dream that has emerged in recent years is not about moving away. It is about coexisting.

Years after I'd read that paper on forest transitions in graduate school, I circled back to Tom Rudel, its lead author, for an interview. He highlighted a couple of factors that make the burgeoning global movement to support a more forested future different from any other known in history. First is the unprecedented level of interest and support. It has all been happening so quickly. Second is the collective action. There are top-down targets, but there is a bottom-up movement underway, what Rudel called "a social movement component."

I like to think of it as a modern-day crisis response, as Mather might have described it, except this one has more people perceiving immediate threats from the changing climate and degradation of forests around the world than ever before. They're aiming for more nature at an unprecedented scale. Perhaps this is a remobilization, born from a perfect blend of humanity's natural instincts for both awe and survival.

As Rudel said, "This is really something new under the sun."

5

A FOREST IS

Tree canopy, or forest cover, seen from a satellite only reveals so much. What those forests are under the canopy and inside the green cover has changed over time.

But what *is* a forest anyway?

That definition matters in distinguishing what people can protect from what they can create. It matters for understanding what our planet is gaining and what it is losing, what the collective dreams for future life on Earth could be.

The first great forested landscape that I experienced outside the United States is now among the last intact, tropical ones remaining. I was seventeen. Volunteering for a program under the Albert Schweitzer Institute during my senior spring of high school, I collected medical supplies from hospitals to send to communities that needed them. I had been reading about Schweitzer's "reverence for life," his ethical view that all life is valuable and important; no life—be it human, animal, or plant—should be sacrificed without determining that a greater good could come from such loss or harm. I was inspired by his service to others, as a doctor, humanitarian, and philosopher. Our medical supplies, along with clothes we also collected, went to Suriname, and I accompanied the last shipment with a small group of student volunteers and teachers. We spent most of our time in the capital, Paramaribo, delivering boxes to schools and orphanages.

If I close my eyes and think back to that trip today, more than twenty years later, most of my memories of Suriname have faded. I vaguely recall walking among the colonial houses along the banks of the Suriname River and helping to paint a concrete wall outside a school with a vibrant mural, of what I no longer know. I remember sweating in the classrooms we visited and my arms sticking to the small wooden desks. I remember listening intently to the beautiful Dutch language of the Surinamese people. But the forests, and my brief time in them, I recall in fine detail—as if their greatness was felt instinctually, their life imprinted upon mine, as if my young mind already knew that our lives were intertwined.

Today, Suriname is the most forested country in the world, with primary forests covering over 90 percent of its vast lands. It falls into the unique category of high forest, low deforestation (HFLD) countries, which have greater than 50 percent tree cover and a deforestation rate of less than 0.22 percent per year.* The HFLD countries, including Gabon, Guyana, Peru, Bhutan, Democratic Republic of Congo, Belize, Colombia, and a handful of others, contain about a quarter of the world's remaining primary forests—those that have reached notable age and remain relatively undisturbed by human activity. These HFLD countries retain some of the "big blocks—our greatest hopes," as Matt Hansen called the relatively large, intact tropical and subtropical forests that remain. Those big blocks are in the Amazon and Congo Basins, in Papua New Guinea, and at the center of Borneo.**

---

* The latter percentage doesn't sound like that much, but the loss becomes substantial when it continues year after year. The term *HFLD* is applied to countries that had greater than 50 percent cover in 2005. The 0.22 percent threshold was selected based on the global average deforestation rate from the reference period (1990–2000), as concern was mounting over tropical forest loss. Even at that time, high-level policy discussions were considering the carbon they stored. How could the countries with biodiverse, standing forests garner financial support to sustain those carbon sinks long term? I loved the title of a paper that identified the need for "preventative" incentives to avoid deforestation. It was called "No Forest Left Behind," also pointing to the inequities surrounding who benefits and how from a standing forest, as well the question of who pays to make treekeeping possible. (Fonseca, Gustavo A. B. da, Carlos Manuel Rodriguez, Guy Midgley, Jonah Busch, Lee Hannah, and Russell A. Mittermeier. "No Forest Left Behind." *PLOS Biology* 5, no. 8 [August 14, 2007]: e216. https://doi.org/10.1371/journal.pbio.0050216.)

** Scientists endeavoring to monitor these blocks defined an intact forest landscape (IFL) as an "unbroken expanse of natural ecosystems within areas of current forest extent, without signs of significant human activity, and having an area of at least 500 km^2." In 2008, a team published an IFL map that represented the first global assessment from high-resolution Landsat data. Scanning the distribution of these IFLs years later, I thought of Jean-François Bastin—"If you cannot map it, you cannot really develop any strategy to conserve it." (Potapov, Peter, Aleksey Yaroshenko, Svetlana Turubanova, Maxim Dubinin, Lars

One afternoon we drove out of the city, where almost half of the country's population resides, and made our way inland from the coast. It was as if we penetrated a wall of green and entered primeval forests with canopies towering above, vines and foliage sweeping down low. I cannot recall if I learned any species names back then, but I remember seeing and touching more than I could have counted if I had tried—tall palms extending stories above, flowering trees with ovate leaves and small white flowers trailing down to the ground in clusters like waterfalls. I remember the leaves dangling around me and crunching beneath my feet; the striking variation in shapes and sizes, textures, and smells of life so diverse.

Somewhere along the narrow and bumpy road, we stopped in a small village for the night. I had trouble falling asleep, feeling hot and sticky while listening to leaves rustling in the breeze. I crawled out from under my mosquito net, which draped over a thin mattress on a creaking frame. I walked outside with a flashlight and got down on my hands and knees to examine the forest floor. The soil was darker than any I'd ever seen, so rich with organic matter, nutrients renewed year after year. I pinched a sample from the surface and rubbed it gently between my fingers. The brown spread further, emitting a refreshing, earthy aroma. I held it up closer to my nose. I picked up a leaf that had fallen from one of the trees above. Thick and leathery to the touch, it was long, wider in the middle, with a point shaped like a lance at one end.

That forest expanded my view of what a forest is and what a forest could be.

In the years that have passed since those few days in one of the most diverse and intact forests on our planet, I've visited only a couple others that rival its vitality. In the Brazilian Amazon and in Borneo, Indonesia, I was drawn similarly to the leaf litter and dark soils and the distinct feeling of existing within layers of life. The branches, foliage, and vines engulfed me in ways I had never experienced as someone who grew up in the eastern United States. Yet, by the time I'd visited those forests, the pressures to

Laestadius, Christoph Thies, Dmitry Aksenov, et al. "Mapping the World's Intact Forest Landscapes by Remote Sensing." *Ecology and Society* 13, no. 2 [2008]: art51. https://doi.org/10.5751/ES-02670-130251.)

deforest were already mounting with the expansion of soy and cattle farms in Brazil and palm oil plantations in Indonesia.

My earliest impressions of what a forest is were first shaped by oaks and maples, big trees in small patches or standing alone in backyards and on small farms. Eastern white oak and chestnut oak, *Acer* species—*saccharum*, the sugar maple, in particular. Bursting autumn colors and the taste of fresh syrup, the imprints of my childhood. When I began my environmental studies in college at Brown University, the narrow streets in Providence, Rhode Island, filled every autumn with dried leaves of only a few species.

In Chile, not long after my pursuit for *la cruz en el árbol*, I encountered my first monoculture forest plantation. I remember feeling such an unexpected sense of confusion. Trees surrounded me; so, technically, I was in a "forest." But all the trees appeared alike, neatly cultivated in rows and evenly spaced. They were just one species of eucalyptus, consistent in height and age. It felt unnatural and manipulated, like a forest imposed upon the Earth rather than one coming from it. It was not the natural and primeval forest that once was but instead one fostered by people's needs—by demand for pulp and paper and a fast-growing product.

Should we really call this a forest too? I'd wondered.

Forest comes from the Latin *foris*, meaning "outside." *Forestis silva* literally means "wood outside." Yet how people define and perceive a forest has evolved from medieval times. There are over 800 definitions of forests in the scientific literature. None of them are poetic or spiritual in nature. Instead of describing the intricacies or the beauty, they're meant to make quantification possible. Before anyone really had an interest in describing all forests across the world together, definitions made sense in place, in context, in different ecosystems, forest types, and land uses. Each either discriminates the land by its human use, such as the boundaries drawn for timber resources, for example, or by its vegetation cover, the biophysical, what literally covers the ground. In our discussions about the extent of the world's forests, Jean-François Bastin had pointed me toward a study from 2016 that determined what was causing a range of estimates of global forest area from satellite-based imagery: it identified "the major reason" underlying the variability as "the ambiguity in the term 'forest.'"[1]

A forest is one or more stands managed together under a single ownership or a single certificate. Such a *stand* is a contiguous area of trees that are relatively homogenous and tend to have a common set of characteristics that distinguish it from adjacent areas. A national forest is a forestland owned and administered by a national government; *forestland* can indicate land reserved for the growth of forests or managed to yield wood and other forest products. In the United Kingdom, a royal forest can refer to wooded areas or preserves that were set aside for royal hunting. Such "kingwoods" are defined more by the intended land use than the presence and composition of actual forest. Choose any one of these definitions, for example, and the selection might exclude some lands that still have forests based on other interpretations of what a forest is.

A forest is a biological community dominated by trees and other woody plants. A closed forest is formed by trees at least five meters tall with their crowns interlocking, covering 65 percent of the sky or more. Many of the biophysical definitions use one threshold or a combination of two different thresholds: the density of trees—how spaced out they are on the landscape—and their height, which relates to age and species. Sometimes, there's a third: the minimum size of the area containing those trees. A forest is an area with high levels of canopy closure—for example, 30 to 40 percent, depending on age, in Russia or 60 percent in South Africa.[2]

The Food and Agriculture Organization of the United Nations uses this definition: a forest is land spanning more than 0.5 hectares with trees higher than five meters and a canopy cover of more than 10 percent, or trees able to reach these thresholds in situ. It does not include land that is predominantly under agricultural or urban land use. "So, that means you attribute the same value to 20 percent, 30 percent, 50 percent, or 100 percent tree cover," Jean-François had mentioned when we talked about how the definition determines what scientists assess as present. "Then you say you have nothing when you have between 0 and 10 percent." (In the Global Potential analysis, he and his colleagues had endeavored to eliminate the threshold approach for this very reason. Even if 1 percent of a pixel was covered by trees, that land area made it into the dataset. It was a novel approach that got lost in the megastorm.) Jean-François had also told me about one of the first meetings he'd attended at the United Nations years ago with representatives of different ministries of forestry; they'd talked about the definition of forest for two days. There's not even one universally accepted definition of a tree. Sometimes the line between a large shrub and a small tree gets blurry too.[3]

In one article about what a forest is, I came across a photo of a verdant valley seen from a road to Ephesus, an ancient city in Turkey. There were a few trees in patches amidst open fields and then rows of freshly planted saplings next to grids of larger individuals; the distant hills appeared covered in forest not organized in any such manner. The caption described the location and then read, "What areas are counted as 'forest'?"[4]

"The definition of forest in Namibia is not the definition of forest in Canada or the definition in Congo," Matt Hansen had explained to me when we spoke that morning of the James Webb Telescope launch. "You'll never get all that data to harmonize from the ground up."

To quantify global forest cover or assess changes, there must be a unified definition of a forest so that it's consistent across the world, and the results of forest cover in one place can be compared to forest cover in another. In remotely sensed imagery, the canopy of the old-growth, the ancient, the primeval, or the primary forest has a distinct texture; shaped by the complexity of the community, the branches intertwining in the sky, the depth of relationships established between long-lived trees. It's like looking down on the contours of broccoli as opposed to a smooth carpet of green that might indicate another forest type.

Confining "forest" to one standardized definition that a satellite sees in data collected around the Earth makes mapping the global coverage possible. But what *are* those forests? One definition might not always capture them all.

Nevertheless, with the advent of global forest monitoring came the birth of a broader but consistent vision of "forest." In the respective studies they led, Matt Hansen and Xiaopeng Song had restricted trees to vegetation over five meters.* Virginia Norwood's sensors on Landsat 1 had

* They couldn't get height from Landsat, but they could use higher-resolution images and other sources to make what scientists call training data for the remotely sensed images. The taller the trees, the greater light extinction due to canopy structure, and this shadowing effect facilitates the forest mapping. In the end, when trees are five meters or taller and sufficiently dense, they are readily mappable from space with time-series imagery and the aid of shadows on the landscape.

delivered pixels at eighty-meter resolution; after data processing, this was resampled to pixels covering sixty-by-sixty-meter squares of Earth. That's a large land area to classify as forest or nonforest. Subsequent satellites and other approaches to capturing remotely sensed data increased the resolution. Matt Hansen and his colleagues had used 143 billion pixels at thirty-meter resolution in the first global analysis of loss and gain—talk about big data! The resolution is still getting better as I write.[5]

Yet some views from space can still miss a lot. Take a mangrove forest, for example—scattered trees on the coast where land meets sea, their roots reach down to the water like baskets turned upside down. A narrow strip of mangroves can get lost in the ocean blue. In their early years, saplings starting to claim their home take years to cross the height cutoff. Some grow faster than others. Do they comprise a forest when they are just beginning? Or is a forest only what they later become?

The line drawn between forest and nonforest is arbitrary, born from the desire to know where all the forests are. In mapping the forests of the world, other small trees—spread far apart in the African woodlands, shrublands, and grasslands—might not get classified as forest. The less dominant signal of those trees is outweighed by what surrounds them in one pixel of planet, meaning it might be classified as nonforest. But if you were walking among those trees, would you see them as a forest?

"It's more inclusive than omitting," Xiaopeng Song told me. But it's also not perfect.

Out of curiosity, I asked a few friends who live in different places what a forest means to them. I love the range of responses from these fellow ecologists and nature lovers:

"A self-sustaining ecosystem that has many varieties of trees."

"Continuous tree shade, layered structure in canopy and understory, diverse in animal and plant species and age."

"Indigenous trees covered in mosses and lichen, creating a home for fungi and shoots."

"Many plant species, including trees, living in relation to each other in multiple states of life and decay."

"Sounds. Smells. Textures. Scales." I could feel the forest.

My father-in-law, a quiet, reflective Brit and philosopher, wrote me this: "'Forest' is a human construct—allowing us to talk about a large collection of trees without naming them all and then to discuss their properties, or the consequences of their destruction."

Ursula Biemann is a Swiss video essayist who reflects on the world and its societies, including the relationships between plants and people. She has collaborated with Inga people in the Amazonian forests of Colombia through both remote work and visits in recent years. When we spoke, I asked her what those definitions based on percentage canopy cover or composition overlook.

"For the Inga or the Amazonian Indigenous," she said, "there is some kind of spirited dimension of everything that is alive." She described how they walk the land and forests to reconnect with themselves, to create links with other beings, to heal the injuries caused by extractive activities. "In the Western world, you think you know something when you can place yourself outside of it and observe it and analyze it and characterize it and measure it and so on. Whereas in the Indigenous world, you know something when you understand its intention, when you understand its actual being, and you understand its place in the ecology that it is forming with the other."

That's a stance that extends far beyond properties or characteristics. The forest is a family, a community of beings.

When Calder was two, my husband Matt and I took him to Glacier National Park for a week, traveling north from Bozeman to nearly the Canadian border. Inside the park, Calder burst into tears at our first stop, a gorgeous lake in a glacially cut valley. We had mistakenly used the word *park* for weeks in advance to build excitement for the trip and realized upon arrival that clearly a "beautiful park" should have swings and slides. His overwhelming disappointment eventually gave way to marvel, and he began wandering around the trees while we cooked. He would toddle up to a tree, put his hand on the bark, and then look back at me, beckoning to introduce me to his new friend. He gave them names—not species names, of course—but names like Douglas, Frederico, and Martha.

As my boy guided me tree to tree, I recalled a passage from Robin Wall Kimmerer's book *Braiding Sweetgrass*. She describes how different languages shape different views of the world; how in English and many other languages we teach "it" for anything in nature. At some point, a child learns that a bird or a tree is just "it." "Names," she writes, "are the way we humans build relations, not only with each other, but with the living world."[6]

I wondered how long Calder's instinct to make friends with trees would last. If he knew more about the forests that have been cleared or converted or just changed forever, would he fight to get them back? As an old friend of mine in Alaska once told me, a healthy forest doesn't always fall into a category with much scientific veracity. It's a feeling.

A fleeting one. Yet more and more people are trying to get it back.

Whereas many scientists working with satellite imagery began their research in one region here or there or came upon the power and beauty of global mapping as technology improved, Matt Hansen, who revolutionized monitoring of the world's forests, has always worked at the global scale, starting in 1995. I can only imagine that he has far surpassed journalist Malcom Gladwell's rule of thumb that attaining true expertise in any skill requires at least 10,000 hours of correct practice. The amount of time Matt has spent assessing changes in forests rivals anyone else's on the planet. Since 2013, he and his colleagues have rolled out updates to the dataset, along with extensive reports from the latest analyses. In the tropics, in the most intact, biodiverse, and carbon-rich forests, such as the Amazon, they monitor in "near real time"; it gives them the ability to flag illegal clearing, or to catch the spine of a new fishbone sliced into the green.

"We did a study on fragmentation of tropical rainforests, and it's basically saying, the smaller the fragment, the much higher probability of it eventually being cleared," Matt told me. "So, that means that if you are in West Africa, Central America, Southeast Asia, where all that is left are fragments, they are going to get cleared." He came across as more clinical and realistic than pessimistic.

Even the "big blocks" must be defined in any attempt to map them or assess them across our world. An intact forest landscape is a seamless mosaic of forest and naturally treeless ecosystems with no remotely detected signs of human activity; it's large enough to maintain all the native species. But a scientist trying to map those intact forests must create some threshold. The line demarcating intact forest landscape is 500 square kilometers; a wooded area of just 499 square kilometers wouldn't count.[7]

Why does *intact* matter? The short answer is that an intact forest offers more benefits than a degraded one. As a lover of trees, I often find myself relishing the shade they provide on a trail as refuge from the summer sun. I savor the feeling that comes with being deep inside a flourishing forest. As a scientist, I am trained to think of that shade or even that feeling as an "ecosystem service," just one example of the many benefits that trees provide to people. With their canopies intricately layered across time and ages and species, with their roots intertwining beneath the ground and connected by a complex network of fungi, the large tracts of primary forest are offering that cool shade and a magical feeling but also so much more.

They can generate rain and reduce the risk of drought. They retain soil moisture and stabilize the water table. They offer habitat for the pollinators that are critical to food production. The list of services goes on; former colleagues of mine at the Wildlife Conservation Society wrote a journal article titled "The Exceptional Value of Intact Forest Ecosystems" detailing their extensive benefits. They can buffer human communities from the effects of extreme climatic events, like heavy rains. Local and regional weather patterns are partly a function of intact forest cover and condition.[8]

"We can't think about all forests being the same," Australian conservation planner Hedley Grantham told me when we spoke about what defines forests, what they have been in the past, and what they are becoming. "We can't think about all coral reefs or grasslands being the same, either. Part of conservation and management is thinking about the differences in biodiversity, and the scientific community has made progress on that, but what we have struggled with is thinking about the condition, or the integrity of those ecosystems."

Ecological integrity is also relevant to what a forest is. Such integrity is about the degree to which a forest—or any ecosystem type—is relatively free from human modification that alters how it is shaped, what it comprises, and how it functions in the world. It's not just intact or nonintact. Integrity offers more insight and detail.

"If you imagine an island of forest, the whole circumference of the island is an edge, but it's a natural one. That's different than an island of forest in an agricultural field that was once all forest." In Hedley's research, a pixel inside each of those forests scores differently when it comes to integrity—because what was there before and what is nearby also matters.[9]

Hedley and his colleagues reported that about 40 percent of forests on Earth are of high integrity. I imagine that first great forest I visited in Suriname scores among the highest. Yet, reflecting on their estimate further, I see it as both a stunning summary of loss and a prospect for hope.

Can humanity find a way to preserve the integrity in the relatively intact 40 percent that still exists and to restore it elsewhere in the other 60 percent, inching the green along a continuum toward some new combination of what it was and what it could become?

PART II

RENOVATION

Patient, plodding, a green skin
growing over whatever winter did to us, a return
to the strange idea of continuous living despite
the mess of us, the hurt, the empty. Fine then,
I'll take it, the tree seems to say, a new slick leaf
unfurling like a fist to an open palm, I'll take it all.

—Ada Limón from "Instructions on Not Giving Up"

6

CHOOSING THE TREES

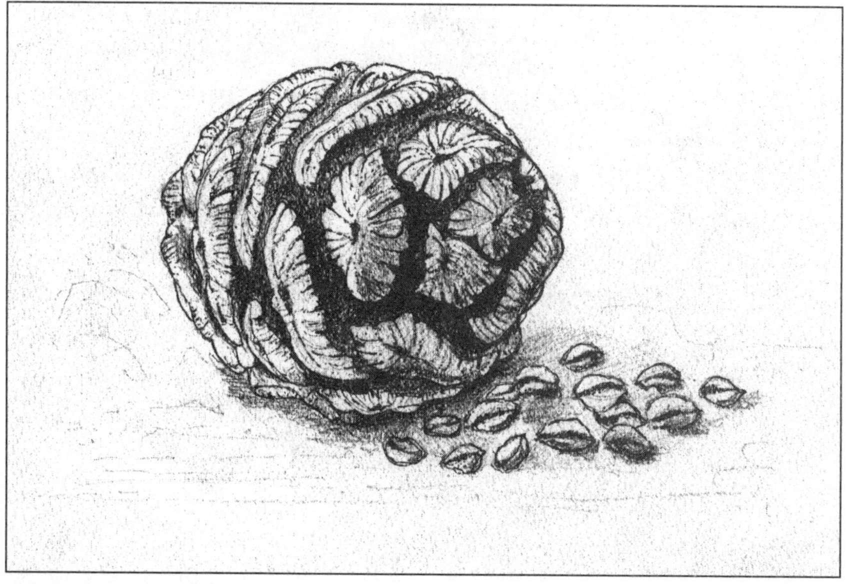

Calder's giant sequoia cone and seeds.

Credit: Lorenzo José Rosenzweig

Most days I write at my desk, a thick piece of plywood that I sanded with care and finished glossy and smooth. Out the window, I see Little Ellis, a forested mountain on the edge of the Gallatin Range. The plywood rests on a sawhorse and a black cabinet. Inside one of the drawers, hidden beneath the pens, pencils, and Post-it Notes, is a little plastic bag with seeds from a giant sequoia. They came from a cone that Calder picked up, unbeknownst to me, sometime during our visit to Three Rivers and Sequoia National Park. I don't know where he found it, but I remember his little pudgy hand clutching that cone on our journey together back home to Montana.

The cone sat in a bowl on my bedside table for a year, maybe longer, offering memories of our time amidst the giants. Then one day, I lifted the cone to hold it in my palm while I was reading, and I noticed seeds had fallen into the bowl. I shook the cone gently, setting the others free, and then collected them all—about 100 seeds in total. I've kept those seeds for years now. They are about the size of a ladybug in diameter, thin, and reddish brown. A dark line runs from one tip to the other like the midrib through a leaf. I thought about trying to plant them in pots or in our garden to see if they might grow far from the original source, but I never did. I kept them as a reminder that people have been moving around seeds and plants for millennia; that a sacred and colossal tree can come from something so tiny; but, also, that a seed is only a possibility of a beginning.

What anyone plants today will have to survive the future to grow big and tall, and that means enduring changing climate conditions like droughts or floods, warming temperatures, cold snaps, and increasing fires. Moreover, the spot where a cone drops or seeds land naturally, without a person or people intervening, may not be a suitable home in the years to come.

That sequoia cone, its seeds, and my swirling thoughts about what and where people choose to plant, as well as what might survive into the future, inspired a trip north. I didn't go to the edges of the Amazon or the deforested areas of the Congo, the tropical hot spots one might expect. I didn't go to Costa Rica or Vietnam or any other country known for a dramatic forest cover reversal. I went to a little town called Kelowna in British Columbia (BC) and then north to Vernon. It was January 2020, and that was one of the last maskless trips I took before the COVID-19 pandemic unfolded.

Within hours of landing in Canada, I was standing inside a freezer that bore no resemblance to any other I'd ever seen. From the outside, it appeared as a massive concrete structure with excessively large doors adjacent to more buildings, similarly prominent on the concrete landscape. I tried to take notes while shivering at the controlled temperature, well below freezing. Stacked high above me, white and brown boxes formed the aisles that I perused.

"This year we're storing 164,696 cartons," announced Sheilagh Fitzpatrick. She'd grown up in a family of fruit ranchers that had then transitioned to the fruit and tree storage industry in the late 1980s. She was the office manager of Hawkeye Holdings Ltd., a family-run business in Kelowna and the largest freezer storage facility in the province. More prepared for the tour than me, she wore a puffy parka that reached her knees. Sheilagh checked the numbers on a piece of paper and proudly declared, "That amounts to 49,183,963 trees."

I had come to British Columbia to learn about the province's progressive effort to get out in front of climate change when it comes to planting trees. It's called *climate-based seed transfer*, and it's far from business as usual. It's an innovative adaptation strategy for promoting healthy forests over time, based on matching seedlings to the projected climate of potential planting sites.

I'd come to see the next generation of white spruce, Douglas fir, and western larch and to understand how and why scientists determine the locations for planting various species and populations. But I'd also traveled there because I was thinking about all the trees that will be planted in coming years, as well as the many other afforestation, reforestation, and restoration efforts I was researching. Choosing what trees to plant predetermines the forest community; it shapes what a forest ultimately becomes, what it offers over time, and whether or not it might endure the impending environmental changes. The storage facility was my first stop.

I ran my frigid fingers across the boxes and investigated their labels. Sheilagh selected one to decipher the codes for me, explaining where the seedlings had come from and where they were going next.

"It's a well-traveled lot," she proclaimed.

From their original sources to their final destinations—through seed collection to nursery cultivation, storage, and planting—this crop would travel more miles in the province than any other had in the past.

"Basically, what we're doing in those boxes is mimicking nature," Sheilagh explained. "They go into the cold. They go to sleep."

Storing seedlings for the winter in a dormant state enables foresters to plant more across the seasons. Planters in British Columbia can start on the coast as soon as spring arrives, move to the southern interior, then work their way north and, eventually, upward in elevation as the snow melts.

"So, these are the forests of tomorrow?" I asked, as she guided me down yet another row. "Exactly," she said with a smile.

British Columbia has been well positioned in many ways to get out in front of climate change in its tree-planting efforts. Since 1987, all timber licenses on publicly owned land have required reforestation after harvest, which perpetuates a cycle of annual plantings and has generated the science to inform action over the years.[1]

Between 1950 and the time of my visit, about eight billion trees had been planted in British Columbia. A government initiative that launched in 2017 was supporting plantings for carbon. Those efforts were typically occurring in areas affected by some sort of natural disturbance, which may not have required reforesting otherwise. Recent fires had created more demand for reforestation. The area that burned in 2017 and 2018, for example, far exceeded capacity to replant.[2]

Alan Rasmussen, seedling and reforestation specialist at BC Timber Sales, later told me that when the agency staff went to order trees in the fall of 2019, the Western Forestry Contractor's Association came back with a limit. If too many trees were ordered, they simply couldn't plant them. "That was the first time that happened," Alan reported. Planting trees across vast lands is physically demanding. A 2021 documentary reported that planters in the province typically burn about 8,000 to 9,000 calories per day; that's like the expenditure of running two marathons.[3]

In the year that I visited the region, about 300 million seedlings awaited planting in the upcoming season. Yet there was still more demand for planting across the landscape than people could tackle. From seed sourcing to collection and nursery cultivation to storage, each planting cycle takes about two years. It's a military-like operation requiring tremendous coordination on a startling scale; the massive freezers played one role in a much larger process.

"It's wonderful when the politicians say, 'We're planting all these trees.'" Sheilagh replied, when I asked about her thoughts on the increasing interest in tree planting. "But hey, where are you storing all those trees?" she added. "Where are they coming from? Who's planting them?"

When we'd finished perusing the aisles of boxes in the first building, she suggested we head to the next building. It felt like a moment of reprieve when I stepped from the subzero freezer into the winter conditions outside. The sun warmed my body, although I could still see my breath.

I wondered about the amount of energy needed to maintain the proper temperatures of those freezers and the other carbon-consuming activities that might be hidden in the planting cycle. Sheilagh said the planters used to pick up their seedlings with an insulated truck and plant them that same day. "The technology hasn't changed much but the distance to planting has." Depending on how far they now go to plant seedlings into their future climate, daily trips just aren't always feasible.

We stood before another concrete building while she fiddled with the door lock. Sheilagh called them "controlled atmosphere buildings," which I learned meant the storage company could suck the air out of them in years when the facility stored fruit instead of trees. A low-oxygen environment slows the ripening of the fruit, so apples, for example, can be stored for months before shipping to a grocery store.[4]

Yet, given the demand for seedlings, there wasn't any fruit stored on site that winter. "Normally, I don't know until September what's coming in November, but everyone knew this was going to be a big year for tree planting. I had to tell the farmers, 'Sorry, we can't store your fruit this year. We're all trees!'"

Where every one of those trees would go had already been determined. Greg O'Neill, a scientist at the Ministry of Forests, was the person who would tell me how.

I met Greg the next morning at the Kalamalka Forestry Centre in Vernon, about a forty-five minute drive from the freezers. His office was small and cozy. On a bulletin board, he had pinned maps of seed source locations and photographs of planting trials. I stared at an aerial image of a peculiar forest. There was a perfect rectangle of trees with dead and dying foliage surrounded by other trees that were flourishing green. Something was affecting the health and survival of the planted population in that little browned area.

"I like to use the terms *safe seed transfer distance* and *migration distance*," Greg said. "If we could rally around those terms, this whole conversation will be clearer."

Trees gradually migrate, altering their range in the landscape at relatively slow rates without human intervention. Through evolution, species adapt over time to become better suited to their local environments. The problem occurs if the environment changes too quickly—as is happening now. When evolution through natural selection and migration can't keep up with the rate of change, some populations die off in areas where they are most vulnerable, or species go extinct. My doctoral research focused on the climate-induced mortality of yellow-cedar trees in Alaska, so I've witnessed the large swathes of centuries-old trees suddenly pushed beyond their threshold for survival. The iconic members of the once-verdant forest turn gray and ghostly as other populations and species that are more suited to the new climate conditions take over and flourish.

"We should expect there to be an evolutionary lag between the climate the trees are ultimately adapted to and the climate that they're experiencing," Greg explained. I think of those trees as being ill prepared or ill equipped for the present and future; they are primed instead for conditions of the past.

Greg relies upon what he refers to as migration distance, an effort to compensate for the evolutionary lag between the climate under which the trees evolved and the climate they'll face in the future. In the practice of planting, it's the climate-informed distance between where the seed came from and where the seedling should go.

"The mean annual temperature in Vernon is about eight degrees [Celsius], but the trees in Vernon evolved in a climate that was six and a half degrees," Greg said. "If you're a forester planting in Vernon, you wouldn't use Vernon seed. You'd go south to Osoyoos, where the climate is nine and a half degrees, to procure your seed because those trees evolved in a temperature of eight degrees."

I'm a scientist in a field quite close to his, and my mind was already twisting, trying to visualize the calculation from the landscape and its climate.

He grabbed a sheet of looseleaf paper and drew two contoured territories in black pen. They overlapped one another, creating distinct areas to the north and south.

"We can collect seed here, but we've lost the ability to plant that seed here," he said, pointing to the area on the southern end. He hatched that territory in one direction to indicate unsuitable conditions under future climate. The challenge is that seedlings planted in that southern portion of the current seed zone would likely do well at establishment but then struggle some sixty or seventy years later, maladapted to temperatures that would then be even warmer. Climate-based seed transfer in British Columbia matches seed from the climate of the mid-twentieth century with the expected climate twenty years later; that's when the planted forest is at one-quarter of its lifespan for commercial forestry. If the time frame for the management goal—timber harvest, in this case—were different, the planting strategy might be as well. Foresters must consider how long they want the trees to thrive into the future.

To make this work, Greg and his colleagues across the province have been drawing upon decades of research—some 1,400 genetic field trials, 550 provenance trials, and a ten-year study of tree species' climate tolerance. *Provenance* means "origin" and in this context refers to a population of trees that come from a particular location. Prior to 1990, the provenance trials focused on finding sources of good seed and assessing

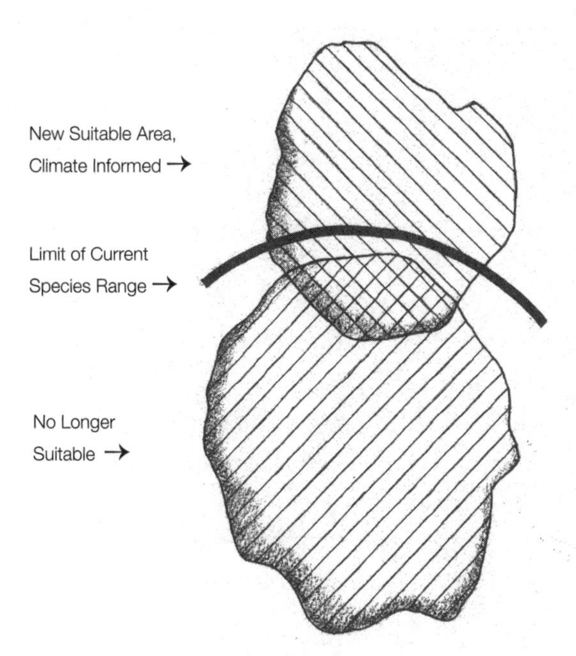

Credit: Lorenzo José Rosenzweig, adapted from Greg O'Neill's original sketch about assisted migration

the impact of transferring seed from one location to another. Those trials proved to be ideal laboratories to test the effects of climate change on seedling establishment and growth, based on the climatic conditions of the seed origins.

He drew an arc that sliced through the planting area in the north. "Sometimes this new area of suitable habitat is *outside* the species range." And that's where moving trees gets more controversial. "The minute you say assisted migration," Greg noted, referring to moving a species beyond its current extent, "well, then there's this outcry of playing God."

Think of assisted tree migration as rungs on a ladder. The first step, assisted population expansion, involves taking seed from the south end of the species' range and transferring it north within its range. That's just moving around populations within the area where the species is currently. Next up is assisted range expansion—moving species outside their current geographic range but still within a climatic one. Finally, exotic translocation goes one step further: moving species across continents or far distances, introducing them to others they haven't interacted with before—as far as we know.

With each step, comfort level can fall. Some ecologists say all options should be on the table and pursued, to some extent, in the "right places." One risk of moving species to places where they don't share an evolutionary history with the other members of the ecological community is that the exotic can spread and push out native species. Exotics can introduce diseases and other pathogens or reduce biodiversity. Eucalypts, for example, are among the most widely planted trees in the world. Yet studies in Portugal and California that compare diversity between "exotic" eucalypt stands and native oak groves show a lower biodiversity of plants, amphibians, and arthropods in the eucalypt forests.[5]

The rungs on the ladder also represent a gradient of perceived risk with respect to possible unintended consequences. I see that gradient as running from yellow to red, as in "generally proceed with caution" or "stop and consider the risks before deciding whether to move forward." A 2018 study led by Guillaume Peterson St-Laurent examined public support for reforestation efforts in British Columbia. The researchers found that most respondents supported or strongly supported assisted migration within range. Opposition increased when considering planting outside range and rose even more for reforesting with nonnative species; such assisted migration "represents a contentious element that is, in the eyes of the public, somewhat comparable to the use of exotic species or GMOs [genetically modified organisms]." A similar pattern was observed in the expert community. Other research has shown those views to be malleable. As climate change forces more people to consider what species and populations will survive into the future, letting go of the firm boundaries between the inside and outside of the current range is becoming more common practice. When Greg and other climate-adaptation specialists use the word *safe* for the seed transfer distance, they're referring to the likelihood that moving those individuals for planting will be successful over time.[6]

"If you put a child in a healthy, thriving environment, the chances of that child becoming a happy and well-adjusted adult are higher," Greg told me. "That's basically what we're trying to do with the trees."

To facilitate the survival of the future fittest, foresters in British Columbia have focused primarily on shifting populations within their current species range, with a couple of exceptions, such as western larch. But new research aims to assess the habitat suitability for species under future climate scenarios. With more information from studies underway and input from Indigenous people living in the region, moving species outside their current range could become a more common practice in the future.* Then local foresters would decide which species to plant, look for the appropriate seed source informed by the migration distance, and go through an approval process for the alternative approach.

I asked Greg how this initiative ensures diversity, given that more diverse ecosystems are more resilient to disturbances. Such diversity comes in many forms—from structural (a mixture of small, medium, and large trees) to compositional (the communities of plants, for example) to genetic. If you're collecting seeds to cultivate and plant, maintaining the

* A new approach to lawmaking processes in BC ensures every ministry and sector of government works with Indigenous peoples in developing provincial laws, policies, and practices. ("New Guidance on Legislation Supports Indigenous Rights." *BC Gov News*, October 24, 2022. https://news.gov.bc.ca/releases/2022IRR0061-001581.)

genetic diversity is critical.* Greg said that their trials have helped ensure this diversity by carefully tracking parents. The number of parents at any orchard supplying seed in British Colombia may be smaller than the number in a naturally forested area, but the genetic diversity of the offspring can be higher. That's because the parents in the orchards originate from many disparate stands. Essentially, the process often mates individuals that would not have mated otherwise.

Greg suggested that we venture outside to see a demonstration project of trees planted near the offices. Standing before rows of interior spruce and lodgepole pine in deep snow together, he explained that the seed sources of the trees spanned about ten degrees of latitude and captured a wide range of mean annual temperatures. It was a small example of what scientists call common garden experiments: indoor or outdoor plantings of species or populations collected from multiple distinct geographic locations, grown together under shared conditions.

Looking at the rows of trees before us, I could easily see that where seed comes from matters. Some species were generally doing better than others. Yet the individuals that had been sourced as seed from warmer regions now towered over the others after only four years of growth.

To the traditional conservationist who aspires to keep species where they are, moving them around proactively is anathema. Keeping them exactly where they are is also futile. Life has always been evolving, migrating, and colonizing new habitat as climate has changed historically, with ice ages coming and going. Human-induced climate change is a new driving force. One plant physiologist told me he got into research on tree-planting strategies "through the inescapable observation that climate-resilient land management is going to require planting new things in new places." That inescapable future is here.

Yet people have been moving seed to other places and planting far from the original source for thousands of years for one reason or another. In Australia, for example, a study combining anthropological research

* The near death of the banana is a good example of what can happen without diversity. Many bananas sold globally in recent years are from one variety: the cavendish. They are clones of each other. Given the lack of genetic diversity in the crop, a fungal disease has spread widely and caused catastrophic losses in the industry.

and genomic data revealed that Aboriginal people were responsible for moving around *Castanospermum australe*, a flowering tree native to the eastern coast. After a careful treatment that removed toxins, the seeds of these "bean trees" could yield a ground meal for food. "Dreaming stories" of tracks were passed across generations to maintain knowledge of physical pathways that Aboriginal people had traversed. The researchers who discovered the species' dispersal followed these physical pathways inland from the coast to the western ranges. The routes that had been shared through song and story revealed the intentional movement of trees along ridgelines to the various inland groves. Back then, people weren't moving them to help the trees survive. They were moving those trees to grow them where they needed them most.[7]

Strange as it seems, changing planting strategies because of climate change may be a modern-day extension of what people have been doing for quite some time. It's the potential scale that probably feels most disconcerting today. In hundreds or thousands of years from now, or even just decades, maybe the dreaming stories of climate scenarios and projections will help explain the pathways marked by our most resilient species.

When I got home from British Columbia, I went back through notes I had taken during my visit to Three Rivers and Sequoia National Park in California. I had written this:

> *Christy at Big Trees Grove:*
> *Climate adapted seeds*
> *But you need to consider the biotic agents as well*
> *Not just about picking the right climate adapted seeds but also considering the interactions*

Christy Brigham, the chief scientist at the Sequoia and Kings Canyon National Parks, had been talking about biotic and abiotic interactions. The biotic factors include interactions between organisms, such as disease, predation, parasitism, and competition among species or within a single species. The abiotic ones are nonliving components of the ecosystem, like water, or temperature, or sunlight. Her point was that foresters and scientists interested in climate-based seed transfer can get laser-focused on climate and then overlook the possibility that a species or population that is most likely to survive in a future climate might face other challenges.

In one study, populations of limber pine known to be resistant to blister rust were more sensitive to drought. Selecting for those more tolerant of drought conditions, in theory, could entail unforeseen consequences with susceptibility to disease, insect infestation, or other pathogens. In the early 2000s trees in a research trial designed to test for the influence of climate on lodgepole pine across British Columbia were severely attacked by *Dothistroma* needle blight, a disease that causes defoliation. The trial site that contained the largest trees had historically experienced some of the wettest weather, but when the weather became even wetter and warmer, those largest trees suffered some of the greatest losses.[8]

Humanity is facing what Christy called a "no-analog future," meaning we are entering uncharted territory. There's a lot of uncertainty as to how the very complex relationships between life and Earth's rapidly changing conditions will play out over time.

Fuckkkk, I had written next in my journal. *Every time we think we have something figured out, we realize there's something else we didn't consider.* I was using *we* to refer to Western scientists, knowing that often some variable overlooked at one point in time becomes unexpectedly and highly relevant in another. What outcomes will arise in the no-analog future from any actions taken today is also uncertain. Alex Woods, a forest health scientist in British Columbia, told me around the time I'd traveled to Vernon, "It sounds great to be cautious; don't move this there or this here. But what we also know is that if we don't do anything differently, what we'd typically plant won't make it." He says we need to do more for the trees to survive. "Just sitting there and being cautious about moving species when we've also been so reckless about degrading our environment and the climate system is problematic. We've kind of screwed up the whole system, so we've got to be more brave."

During my time with Greg, I'd learned about giant sequoia trees that were growing in an experimental site on Vancouver Island, far north of their current range in California. Planted horticulturally, they can be found thriving in green spaces in the city of Vancouver too. From various scientists whom I later interviewed, I learned about giant sequoias growing in botanical gardens in France and Scotland. Some of the tallest giant sequoias outside the California groves now line the Avenue of the Giant Redwoods in Argyll, Scotland. They were planted in the late 1800s along the entrance to what was then a private estate. A quick online search revealed a library of photographs from around the world—tall sequoias in botanical gardens from Christchurch, New Zealand, to little saplings in pots in Hiroshima, Japan. There was a long list of sightings throughout Europe: Greece, Spain, Sweden, Norway to name a few. People are so captivated by such a long-lived and stunning species that they've carted their cones and seeds all over the place—just as Calder had done. But those planted individuals are scattered in various places across the world "in captivity" as some ecologists might say, not really comprising a "forest" or thriving in the "wild."*[9]

The natural home territory of giant sequoias is quite limited. They occur on the western slope of the Sierra Nevada in a very specific elevational band—from about 4,700 to 7,000 feet. They're restricted to about seventy-five groves, although a *grove* is another human construct without a perfect definition. Groves are essentially small groups of trees that are close together, but there are gaps between those clusters. Giant sequoias grow in wet microclimates that are expected to get hotter and drier in the warming world. The high elevation band where these trees exist in their native range is typically above the snowline. Yet during the 2012–2016 drought in California, some of the groves didn't accumulate any snow at all.

Calder's giant sequoia seeds from one of those trees stayed tucked away in my drawer for a couple years after my trip to British Columbia.

---

* A study published in March 2024 in *Royal Society Open Science* described growing interest in planting both giant sequoias and closely related coastal redwoods in the United Kingdom "partly due to their carbon sequestration potential and also their undoubted public appeal." It reported an estimated half a million of these two species in the United Kingdom, and various news outlets at that time reported fewer than 80,000 giant sequoias living in their native California range. The study found giant sequoias to be generally well adapted to conditions in the United Kingdom and suggested their growth indicates "they are likely to be a good choice [for planting] purely from the perspective of carbon uptake." (Holland, Ross, Guilherme Castro, Cecilia Chavana-Bryant, Ron Levy, Justin Moat, Thomas Robson, Tim Wilkinson, et al. "Giant Sequoia [*Sequoiadendron giganteum*] in the UK: Carbon Storage Potential and Growth Rates." *Royal Society Open Science* 11, no. 3 [2024]: 230603. https://doi.org/10.1098/rsos.230603.)

Then in the summers of 2021 and 2022, I pulled them out again when I was reading more about the fires. The 2020 Castle Fire burned more than 170,000 acres before it was contained, including about 9,530 acres of giant sequoia groves on National Park, US Forest Service, and other public and private lands. Within Sequoia National Park, where I had traveled with Calder, three groves experienced high-severity fires; nearly 375 large giant sequoias were killed in those areas. A complex of fires in 2021 burned another sixteen groves, almost entirely on national park lands. The 2022 Washburn fire in Yosemite National Park hit Mariposa grove, the largest and most popular of the park's three giant sequoia clusters.[10]

Yet planting trees isn't an activity that the National Park Service has historically championed on its protected lands. It has a mandate to conserve park resources and provide for their use and enjoyment "in such a manner and by such means as will leave them unimpaired" for future generations.[11]

Holding Calder's seeds in the palm of my hand, I wondered what would regenerate on the land where those sequoias had burned. If spindly seedlings popped up in the sunlight and freshly exposed soils—as some had done at earlier burn sites already—could they endure a future of warming temperatures and drought? If anyone were to select seed and replant sequoias, would some offspring from certain groves fare better than others? Or like the individuals selected for climate-based seed transfer in Canada, could their best possible future be somewhere else?

In 2022, years after my visit to Sequoia National Park, I called Christy on a smokey day in California. A helicopter had made multiple trips that morning to dump what ultimately amounted to 60,000 gallons of water on a burning sequoia. I gasped at the quantity; that amount of water would fill a tank twenty-five feet in diameter and eighteen feet tall. It hardly seemed like sustainable effort, but the alternative of letting fires run rampant doesn't fly either.

"What's happening now is we have major fires that have killed 100 percent of the tree cover," Christy began when I asked her about the sequoias and whether they were considering any replanting in the park. We were talking during her lunch hour, which she had used for our video meeting instead of eating. She spoke rapidly, riddled with concern and carrying a sense of responsibility for the trees.

"So, you might think that's the perfect environment to experiment with moving things, but the reality is that the Park Service does not have a strong history of reforestation, period. Getting support for reforestation at all has been difficult, and then trying to overlay on top of that 'Oh let's also think about climate change adaptation and maybe we should actually move the grove up 1,000 feet in elevation' seems too difficult at this point."

Christy said it was in her work plan to propose the least-risky strategy for replanting. I thought of the gradient from yellow to red for perceived risk associated with assisted migration; this sounded like yellow—proceed with caution. "We think the most likely future—and what we're already seeing—is hotter, drier conditions. So, we're going to bet-hedge." If approved for planting, they wanted to use the local seed material and add some from drier sites that are further south and lower in elevation. Then they'd mix in other seed from more diverse populations. "We're giving it a kick in terms of the material that the site will have to work with as adaptation and evolution occur," she told me.

"When you say 'work plan,' are you referring to your dreams and desires or something that's actually going to happen?" I asked.

"Somewhere in between those two," she replied, then got up to fetch a report from one of the many packed bookshelves in her office. It looked about an inch or so thick, and Christy clarified that it was only the summary; the full thing spanned multiple binders. "It's called 'A Climate-Smart Resource Stewardship Strategy for Sequoia and Kings Canyon National Parks,'" she declared. "In here for giant sequoias, it says, 'Experiment with assisted migration.' So, that means it's more than my hopes and dreams."[12]

Like the scientists and foresters in British Columbia, Christy and her colleagues had some solid data to inform decision making. Recent studies had shown genetically distinct northern and extreme southern populations. Some groves are more genetically diverse than others.[13]

"We can sit and keep debating what to do? Or just give it a try?" I probed. Christy replied,

> Where we go wrong is to say, "I'm smart, I did a bunch of modeling of what the future climate is going to look like. I'm very confident, and I know it's going to be drought, so I'm going to the south, and I'm only sourcing from there." That's super confident and doesn't acknowledge the mess and complexity. Other people would say, "Well, in the face of uncertainty, don't plant at all or only plant seeds from the local population. Don't assume to know better than nature." I want to get in there and help, but I think, do it messily and be humble. So, get some wetter tree material, get some drier tree material. Use a lot of the material that's already there because it's done well up until this point.

The amount of effort that had already gone into considering the future of these relatively small patches of forest and the prospects for this magnificent species made sense to me. They're storing a lot of carbon and attracting over a million visitors a year to walk beneath their canopies in wonder. Christy called the giant sequoias "the ambassadors of all trees." I'd like to see a next generation reach toward the sky too. What I'd learned from Greg and Christy made me concerned about all the other species that have been planted elsewhere or will be planted around the world and their prospects for survival.[14]

There are indeed similar efforts underway in other places. In the United Kingdom, foresters are moving oaks and beach northward, having already realized that if they plant where they have in the past, the trees will not survive. Some seed even comes across borders from France as a climate-adapted planting strategy. In the western United States, researchers have developed an online platform that helps land managers identify seed sources best adapted to the local climate conditions at planting sites; in California, that voluntary tool operates a lot like the approach in British Columbia to estimating the safe seed transfer distance or migration distance.[15]

However, not all tree-planting initiatives around the world will have the benefit of so much data to inform planting strategies. Extensive networks of orchards for sourcing seed, the ability to collect seed across large distances, and studies that illuminate which populations or species might fare better than others may also not be available. Political boundaries don't always match ecological ones, which are also in flux. Whether regulations align with the intent or desire for mass movement is another consideration.

A former colleague of mine who has worked on reforestation efforts in Asia and Africa told me that the first hurdle in many projects is often just getting landowners or farmers to agree to plant trees on their properties. If they're using the land for agriculture, for instance, planting trees might mean losing that economically productive use. The farmers need to earn an income from those trees, create some added value, or receive some other direct benefits. He said that it can be a real struggle to get them to support planting in general. Introducing the question of *what* to plant under future climate scenarios requires another effort. So, calculating and implementing safe transfer distances on top of all that isn't often practical.

Based in Arizona, Tom Whitham is an expert in community ecology and community genetics. He's been running common garden experiments on select species since the 1980s to understand whether variation between trees in "the wild" was based on environmental conditions or genetic differences. The work wasn't motivated by climate change back then, but it was easy to transition to that focus. He'd taken multiple species of willow and cottonwood trees and planted them all at an experimental site to see which ones proved more resilient to warmer, drier conditions. Fremont cottonwood was very sensitive; he said that coyote willow "didn't seem to care."[16]

> Every time you plant something out, it should be tagged, and you should know where it came from, because you're spending a ton of money just to do the work, but you're not learning anything from it. That's commonly what I find when I go to restoration sites: people plant, but they don't continue to monitor, they don't quantify performance, and they don't know where the seed or seedlings originated. If you have great success, well, then, wow, it would be helpful to know what you used. Where did it come from? And if you had a complete failure, why did it fail?

Tom noted the differences in practices between tree planting and farming. Seed producers in agriculture run trials every year to see what seed will perform best; it's standard practice.

"We're talking about a world in climate crisis!" he told me, as we took a step back to talk about the global restoration movement. "If you're planting millions of trees—maybe even a trillion—without knowing where they're coming from, I'd say you're wasting your money."

Trees take much longer to grow than a season's vegetables, so waiting on multiyear or decade-long trials isn't entirely practical either when the warming world needs positive action now. But I took his comment to suggest that more people need to raise the bar for how they decide what to plant, as well as how they track the outcomes of their effort over time and share lessons learned. The act of planting a tree should be treated as if more life depends on the outcome, because it does.

Calder's giant sequoia seeds now rest inside a little paper envelope on the windowsill above my desk. When I told Christy about them on the phone, she smiled and instructed me to take them out of the plastic bag immediately.

"They need to breathe!" she exclaimed, worried they hadn't dried out enough for proper storage.

They are now my constant reminder to keep asking, "Where did that seed, seedling, or sapling come from? Where is it going?"

7
EXPECTATIONS

In 2012, an elderly woman in the Aragon community of northeastern Spain took it upon herself to restore a 1930s fresco painting of Jesus Christ called *Ecce Homo* by the artist Elías García Martínez. The *New York Times* reported that the woman, Cecilia Giménez, "said on Spanish national television that she had tried to restore the fresco, which she called her favorite local representation of Jesus, because she was upset that parts of it had flaked off due to moisture on the church's walls." In the hands of an amateur artist, the makeover went awry. The result, in critics' view, was terrible; some deemed it the "worst restoration ever."[1]

Giménez claimed she hadn't finished her work; her initial strokes had been discovered while she was on holiday. The portrait had been transformed into a "cartoon smudge, with beady misaligned eyes, and a halo of hair bearing an unfortunate resemblance to a chimp." *Ecce Homo* ("behold the man") became known as *Ecce Mono* ("behold the monkey"), and soon pictures of the "beast Christ" or the "potato Jesus" spread across the internet. Parodies showed the original next to the botched restoration and also beside a Sesame Street character. Others mocked its monkeylike appearance and the melting face. A Facebook group titled Beast Jesus Restoration Society was created and garnered thousands of followers in a few weeks.[2]

Most notably, it caught so much attention because it didn't meet people's expectations of what restoration is or should be—a return of something back to its original state.

I missed the memes and various images that went viral at the time and instead came across a mention of *Ecce Mono* years later in an ecological restoration journal. The commentary, coauthored by environmental writer Emma Marris, described a formula that people tend to expect in any kind of restoration: A→B→A′, where A is the original state, B is the degraded state, and A′ is the restored state.[3]

"Restorations that take the object to a novel state," it argued, "(as in, A→B→C) are generally thought of as failures." *Ecce Mono* offered a suitable anecdote, given the same holds true in ecology as it does in art.

Oddly enough, *Ecce Mono* had brought some unexpected benefits in the years that had passed. One article I came across called it "the first known economic miracle performed by a meme." *Ecce Mono* had resurrected tourism in the small town of Borja, where it still hangs, as travelers flocked to see the rendering. In about a two-year period after the makeover, more than 150,000 tourists came from around the world to see the result. Visitors paid a euro to the church for entrance, ate in local restaurants, and picked up merchandise such as mugs and T-shirts. Lonely Planet, the travel guide publisher, highlighted the painting in a book describing "360 extraordinary places you never knew existed and where to find them." Giménez, who had received much criticism and ridicule, eventually became celebrated annually in Borja on August 25, the day of the transfiguration. Like Madonna—so widely recognized by a single name—Cecilia Giménez is now known simply as "Cecilia."[4]

As the authors of the ecological restoration article explained, the feasibility of restoring ecosystems—including forests—to historical states is declining worldwide. Increasingly people are implementing projects that aim to transform a site into something different than it was previously. The authors highlighted the need for renaming restoration (the ecological kind), because it sets up a certain expectation for what the outcome can or should be.

In my opinion, Suzanne Prober, an ecologist with the Commonwealth Scientific and Industrial Research Organisation, Australia's national science agency, has captured this thinking best. In 2019, she and a team of scientists proposed the use of a new term: *renovation*. They said this term could describe ecological management and nature conservation that actively allows for change and accommodates characteristics that differ from historical states.[5]

I looked up various definitions online:

Renovate *(verb).*
To impart new vigor.
To remodel.
To revive.
To repair and improve something.

With just one word, I felt completely liberated from the unrealistic expectation of re-creating an exact replica of some ecological past in this rapidly changing world. I loved the sentiment and vision of repairing degraded forests, of improving upon the current state.

Not long after my trip to British Columbia, I spoke with Suzanne. She told me she wasn't sure how much the word itself had been taken up, but the concept was resonating and gaining traction more broadly.

"The term *restoration*," she explained, "has a slightly out-of-date feeling for me. Some people want to change the meaning of the word *restoration*. But ultimately, what we need is more people to recognize the need for finding balance between the old and the new." She was referring to the old paradigm in ecological restoration of restoring an ecosystem to what it was with the new reality of what is actually possible.

"In the past, in terms of restoration, we haven't had to make hard decisions about what we're aiming for. It could be guided by what once was," she said. "Now, we have to make hard decisions about what we're working toward, what we actually want to keep trying to sustain of what we value in those original systems, even if we can't have them in the future exactly as they were."

Dying trees could trigger different action later, but people need to make decisions now. "When we are investing in planting trees, for example, it's going to be 100 years before they turn into habitat trees, so we have to get it right." *The same standard*, I thought, *is true when targeting carbon benefits or any others.* If people are counting on the trees planted to continue drawing down carbon from the atmosphere over decades, getting "the what" and "the how" right at the beginning is also critical. That may not be A→B→A′.

Suzanne told me that she and her colleagues chose the word *renovation* because it combines the need for change with a respect for what was there before. I was on board with all her arguments, and I still found it difficult to use the word in practice with other ecologists or to put it into practice in my reporting on the topic. Yet the new framing effectively altered how I perceive restoration and what expectations of it might be. I keep using *restoration*, knowing that A→B→A′ doesn't bring us all the way back to A, the original state.

I still don't think *renovation* accurately describes Cecilia's work on *Ecce Homo*. But I did appreciate the lesson that even an incomplete or botched effort can deliver unexpected benefits—that something different and good can result from an attempt to improve upon a current degraded state.

I've walked with Calder on trails through what I know to be secondary forest; I can see it in the structure. Forests that burned and then regenerated. Forests that were cut and then replanted, the trees surrounding us consistent in age and height and often limited in diversity too. The green is shaped and formed by a new community brought together, but wonder still forms the foundation of his experience. He touches the leaves. He picks up sticks. He marvels at the light. He finds stillness. He simply *feels* a forest. Almost uncontrollably, I draw upon a different knowledge and history. I am remembering and imagining A, the original, what was or might have been, but he is fully present in what C has to offer.

Maybe there's a glimmer of hope in there. What people and nature renovate together can become beautiful in its own right, never fully replacing what was but still delivering on much that sustains life.

8

COEXISTING WITH CONCRETE

When I was in college in Providence, Rhode Island, I made a photo essay about urban trees for an art class. It was less about the trees themselves and more about the space people give them in cities and the help they need to grow. I photographed squares of earth spared from concrete to make room for street trees. I remember questioning the appropriate size of those patches: What would be enough? And was "enough" all they would get? I photographed roots busting through sidewalks, a visual commentary on the strength and persistence of nature and the outcomes of poor planning and long-term care. I photographed stakes, tethers, lines offering support, and trunks enclosed in wire or wooden slats for protection. The essay wasn't a criticism or a mockery, and very few people ever saw those images anyway. In my mind, it was a celebration of the people not pictured—whoever was behind each "scene," working to bring the green back to a place where trees had been forgotten.

In the autumn of 2022, on a chilly morning in downtown Seattle, Washington, I got to meet a lot of those people. I had been remotely interviewing foresters working in cities around the country when I discovered a strange trend: they were all unavailable for calls over the same dates. I inquired for more details and learned they were heading to the Arbor Day Foundation's Partners in Community Forestry Conference. It's an annual multiday affair, and that year's conference had the largest registration and attendance in history. The Arbor Day Foundation has been dedicated to planting trees since its founding in 1972, but as with many other organizations involved in the movement, its capacity and reach have both expanded in recent years. I registered to go too.

If you want to focus on the sheer quantity of people in direct proximity to trees who benefit from their presence in many ways—beyond the carbon they sequester—the urban context is one to consider. The state of California, where I spent many years as a graduate student, has 130,000 acres of public school land that serves about 5.9 million children. That's more annual visitation to school lands primarily in urban areas than seen at Yosemite, where the Mariposa Grove of Giant Sequoias and iconic Half Dome attract people from near and far. In the United States, nearly 85 percent of people live in urban areas, hundreds of which have populations of over 100,000 people. Globally, about 4.4 billion people live in cities; that number is rising quickly. From a city to a remote forest, trees in any place can benefit the global climate system to varying extents. I thought it would be remiss to let my own preconceptions about what a forest is rule out the consideration of city trees in my investigation of this growing movement.[1]

The urban forest encompasses the trees and shrubs in an urban area—individual trees and green spaces with trees, even the vegetation associated with trees and the soil beneath the trees. Some people see it more broadly as the whole urban expanse of concrete and trees, adding yet another dimension to what a forest is or could be. I figured that efforts to increase forest cover in cities might illuminate more motivations at play.

I rode the escalator up to the conference welcome area with two members of the Blackfeet Nation from Browning, Montana. Around the registration booth, women in business skirts and plenty of heels, fancy flats, and loafers contrasted with the usual sneakers and hiking shoes. Yet, it wasn't until I entered the main ballroom, which was packed with hundreds of people, that I could finally soak in the diversity of the crowd. Men and women. Black and white. Indigenous and other people of color. The seats at the roundtables, which were organized in a tight grid, were almost all taken. People stood along the walls. Trees—city trees, specifically—had brought everyone together.

I sat down in one of the last seats. The man next to me immediately introduced himself.

"Kiley Miller," he said, extending his hand for a shake. "I'm with Trees Forever in Iowa," he declared. When I told him I was a writer, he reached into his bag and presented me a report about his organization's "ReLeaf" initiative (not Relief, although I'm sure that was implied). It had a big, green oak leaf on the cover, overlaid with the text "Plan to Bring Back Our Trees." I thumbed through to see planting guidelines, images of treeless streets and broad canopies offering shade, and lists of "superior" species to consider, like white oak (*Quercus alba*), butternut hickory (*Carya cordiformis*), and black walnut (*Juglans nigra*), as well as "allowed" species like quaking aspen (*Populus tremuloides*) and ginkgo (*Ginkgo biloba*).

As I looked around the room, I thought back to those photographs I'd taken in college. I was surrounded by the people I'd only imagined back then—the hidden champions of backyard trees and street trees, of clusters in schoolyards and city parks. Just like the governments, companies, organizations, and communities wanting to restore degraded forests in rural or remoter parts of the world, they were chasing a similar goal: more forest. As the opening talks began, I quickly learned that their work was less about the number of trees planted and more about the number of people benefitting from them in so many ways.

Sure, there's the aesthetic appeal of a city street lined with sycamores, cherry, or birch, but urban trees are far from just decorative. They offer shade and cool the air around them through evapotranspiration, which is kind of like sweating all day: they absorb heat and release water vapor through their leaves. They play an outsized role in regulating neighborhood temperatures, which is becoming more important as cities across the world get hotter.

Relative to nonurban areas, cities in a similar geography are disproportionately warmer. The cause is what scientists call the urban heat island effect: surfaces such as pavement absorb solar radiation and warm their surroundings by reemitting that radiation as heat. Data collected around the world over time reveals the increases in *global average temperature* as a result of climate change; the localized effects can be much more dramatic—especially in cities. A "well-canopied" neighborhood in Portland, for example, may be 20°F cooler on a hot summer day than a lower-income suburb with relatively little tree canopy. Such dramatic differences in temperature also translate into economic inequities. People in urban areas with more trees tend to use less air conditioning and save on energy costs. Trees growing in these urban heat islands can slow the rate of warming and help create stabler microclimate conditions.[2]

They can also reduce the likelihood or intensity of flash floods. They improve air quality by intercepting airborne particles, which they absorb into their tissues or retain on the outsides of their leaves, branches, and trunks. A study of the city of Chicago's 801 census tracts found a 10 percent increase in tree canopy cover was linked to about a 10 percent decrease in crimes such as assault, battery, and robbery, after accounting for other potential explanations. Other studies support converging results across multiple cities. More trees invite people to enjoy being outside and create community through time together. Perhaps there are "more eyes on the street" that discourage crime.[3]

In his enthusiastic welcome, the CEO of the Arbor Day Foundation called this moment a watershed for urban and community forestry—turning to trees as a solution to some the biggest issues facing communities today.

"It's not just the United States. It's Canada. It's Africa. It's Asia. It's all around the world. People are recognizing that it's time for trees, and it's time to be bold. Now we know, it's not about just planting trees and walking away. It's about quality." He declared that in this moment, "we're going to have to push scale like we've never been able to do before and do quality and quantity." The room exploded in applause. Having spent a lot of time in one-on-one conversation with urban treekeepers, I felt like I'd come to the epicenter of action.

Beattra Wilson, the assistant director of urban and community forestry at the United States Forest Service, was up next. As she walked on stage with grace, people cheered and hollered, as if a rock star now stood before us. I felt a bit unprepared, not knowing the scope of her work. Kiley leaned over to fill me in.

Under her leadership, the Urban and Community Forestry Program's funding had surged upward from $32 million, creating unprecedented opportunities for action. "She has about $1.5 billion dollars in federal funding right now," he said, letting the billion roll out slowly for emphasis. I had no idea that level of commitment was in place for urban trees, and I was eager to learn more. In a ten-year investment, the money, which had been allocated through President Joe Biden's signing of the Inflation Reduction Act, was targeting neighborhoods that lack green infrastructure and bear the brunt of climate change.

Later, when I searched for more information about her online, I discovered she grew up in a small Louisiana town and is a first-generation college student in her family. Her exposure to agriculture and land use through 4-H, along with her drive for a career that brought her to a big city, had directed her path to urban forestry. When she began her federal career, her dream was to work at the White House one day. By 2020, she'd completed two rotations.[4]

In good humor, she told the audience, "I went preacher on y'all and created an acronym" for the theme of the conference: BOLD.

At first, Beattra said she'd thought that *B* could easily represent the "billion reasons" she had to be there; then she'd settled on *Be Better.*

"*O*, I came up with *Outrageous* because we need to really be disruptive and go for it and use these resources in a way that people never thought trees could be part of the solutions."

The *L* was for *Listening* and *Learning*—for realizing that trees are the "load-bearing walls for this house we call home, which allows community."

The *D*? *Don't Hesitate.* The time for change, for more trees, for all the benefits that come with them, is now. The money, she noted, is for scaling up.

"As someone who is celebrating twenty-two years of my career in forestry, I am most excited about the diversity of this room I am standing in front of," she declared. In her years of coming to these conferences, Beattra said, she had often been the quiet person in the back of the room, "always mistaken for a student, or an intern, still mistaken for a student or an intern!" The crowd roared with laughter. "But it's so exciting to come into a room that looks more and more like the communities that we have been called to assist, and to bring better and more improved urban forest conditions to."

She was referencing another mission that extends far beyond the number of trees planted, the amount of carbon they could sequester, or even the number of people directly benefitting from trees. She was talking about achieving tree equity.

On June 22, 2021, American Forests, the oldest national conservation organization in the United States, released the first nationwide tally of Tree Equity Scores (TESs); it was an effort to assess the equitable distribution of tree cover across the country. Along with trees, the researchers looked at population density, income, employment, race, measures of public health, and surface temperatures. The TES identified the cities that could gain the most—in terms of health, economics, and climate—by increasing tree canopy in places of high need. The pattern revealed by the national assessment was stunningly clear: tree presence in the United

States strongly correlates with wealth and race. Results also showed the country needs about 500 million more trees planted in the right places to achieve tree equity.[5]

Chris David, vice president of geographic information systems and data science at American Forests, led the TES mapping effort. "The way we address equity," he explained to me on a video call, "is we prioritize where you should be planting, based on what we call our priority index." The index helps identify the places that have been historically ignored, based on a suite of factors. "They often have a high percentage of people of color and/or are low-income neighborhoods. So, we are identifying those areas and prioritizing those above equally deficient areas in terms of the number of trees."

Urban neighborhoods with low tree cover also commonly carry a legacy of redlining, which refers to racial discrimination in housing. Troubling outcomes still exist today from a policy established by the federal government in the 1930s that created barriers to homeownership, employment, and access to education for people of color. The Federal Home Loan Bank developed a program to appraise real estate risk, and it used four classes for levels of security for investments: A (green)—best; B (blue)—still desirable; C (yellow)—declining; and D (red)—hazardous (hence the term *redlining*). Matthew Lasner, an associate professor of urban studies and planning at Hunter College, told the *New York Times* that the neighborhoods redlined by the government varied in all sorts of ways—age of the homes, average home values, proximity to industrial areas—but they typically had one thing in common: black people lived there. Many people have studied the lasting effects of these policies, but only recently has interest in associated canopy cover arisen. A study of urban tree canopy cover in thirty-seven US cities, for example, revealed that areas formerly graded A had nearly twice as much canopy cover (~43 percent) as areas that were formerly graded D (~23 percent).[6]

Those green lines were drawn long ago.

Chris and his team at American Forests found that areas with the most people of color had 45 percent less tree canopy cover than areas with "the most white people" (more than 80 percent white) in the nationwide analysis; the poorest versus the wealthiest quintiles had 36 percent tree canopy discrepancy. As the organization's nationwide coverage later expanded to more locations, these discrepancies in tree cover were updated to 40 percent and 30 percent, respectively.[7]

"It isn't fair," he said. "That's not the way it *should* be, and it doesn't *need* to be that way." The patterns of inequities extend beyond the urban landscapes: in many rural places, people clear trees to gain from more economically productive land use. I thought of Virginia Norwood's insight in 1972. She had been preparing a speech for one of the top brass at the Hughes Aircraft Company to give about the multispectral scanner system sensor, which she had designed, before the satellite went into space. "Census experts like to keep track of the rate at which wild and agricultural land becomes urbanized," she had written. "A side effect here is that the amount of greenery is a good index of the economic level of the neighborhoods. Even on a 200-foot scale the infrared return from Bel Air has a very different character from East Los Angeles."

Chris told me, "Trees are a fundamental right for everybody." I hadn't ever thought about them that way, but I could see the logic.

They are the load-bearing walls of this house we call home.

Before attending the conference in Seattle, I had talked with other urban foresters, volunteer organizers, and data scientists working in cities across the United States to address tree equity.

The American Forests TES analysis offered the first nationwide analysis, but cities like Chicago and Tacoma, which I had planned to visit after the conference, have been embracing similar efforts to unpack and address the intertwined inequities. A mapping specialist and data analyst for the Chicago Region Trees Initiative told me she had put together maps from various datasets in Chicago, and each one looked the same. "If you were looking at temperature, at asthma, at heart disease, at canopy cover, any of these things, they were all poverty maps." There were correlations between them all. "It's all tangled up, and what I saw made me want to work at ways that we can do better."

These overlapping maps that the analyst at the Chicago Region Trees Initiative had referenced do not indicate causation, meaning they're not proving that lack of canopy cover causes heart disease, for example. A study in Tampa, Florida, found the percentage of tree cover to be inversely related to cardiovascular and respiratory health outcomes. Researchers have documented a similar relationship between tree cover and depression among elderly residents living in US nursing homes. A team of scientists that included another colleague of mine from graduate school, Greg Bratman, found an association between tree and shrub cover around urban schools in California and test scores for elementary school kids. These associations illustrate other inequities; teasing out the causation in every case is not easy. Yet clearly the presence or absence of trees is tied to so much more.[8]

In my research, I'd also come across urban resistance to tree plantings. A study in Detroit, Michigan, for example, revealed 24 percent of residents offered a street tree between 2011 and 2014 submitted a "no-tree" request. Data collected from city residents who were eligible for trees, as well as from staff, board members, and volunteers at a nonprofit organization leading the endeavor, revealed that perceptions of neighborhood character and concerns about maintenance were major hurdles to uptake. As I reviewed the details, this response struck me: "You know I love trees actually, it makes things beautiful and it makes shade. There's no problem with trees. The problem with trees is they have to be maintained and in our neighborhood, they're not being maintained. . . . I tried to get the city to trim trees [for] almost 15 years and they never came out." Some residents were more upset about other problems that they couldn't control in their communities. So, a neglected tree or a new tree might just a be a focal point for wider concerns. "It's about being able to control your environment," one participant said.[9]

A Bloomberg article mentioned a "landmark report" conducted by a University of Michigan environmental sociologist in 2014; the report warned of the "arrogance" of white environmentalists when they introduced greening efforts into communities of color. I could see parallels to criticisms of the tree-planting movement in rural landscapes and the Global South as well—this idea that policymakers create a target for a certain number of trees, funders come forward, and then there's this top-down approach of "Oh, hey, this is where the trees could go." It won't work without local support.[10]

The many urban foresters I interviewed didn't talk about planting resistance as a concern. Some called the resistance a persistent "urban myth," given how much has evolved already in how people come together to do this work. In the ways they described their initiatives in various cities, the issue had as much, if not more, to do with engaging with community members to facilitate local ownership, care, and trust across partnerships as it did with reforesting practicalities.

Chris said that he'd expected the TES results to be used for fundraising to support efforts to increase urban canopy cover in priority places. Pretty quickly, cities started using the equity score too. In Phoenix, the city council passed a resolution to pursue planting based on tree equity, and the TES has served as a planning tool. An organization in Rhode Island used the scoring to prioritize tree giveaway programs in certain cities, ensuring trees went to neighborhoods with the hottest temperatures, fewest trees, and other socioeconomic characteristics.[11]

At the Seattle conference, I joined a session on tree equity and a discussion group afterward. My little group of experts included an urban forester and tree equity planner in Columbus, Ohio, someone who worked in Las Vegas (not exactly what I think of as forest), and a representative from the City of Detroit.

The woman from Detroit said the maps of tree equity had changed the conversation. "We don't go up to someone and just offer a tree anymore. We say, 'Here's the heat island problem. Here are the asthma issues. Now here is a solution. Plant this and care for this.' You don't really have to explain an actual score; you can just look at the map and ask, Does this seem fair to you?"

I found myself continually going back to the word *fair.* Not just fair tree cover. Fairness in the distribution of what all forests, including urban ones, have to offer. Fairness in who and what have lost the most already. Fairness in the benefits that people may *feel* among trees but that Western science is only beginning to reveal.

As in the days when I had carried a camera around Providence, I became more acutely attuned to leaves on the sidewalks in Seattle, to individual trees on the streets, and to small patches of urban green. I stopped to admire the sheer size of the big leaf maple foliage at my feet. It was autumn, and the colors took me out of my head and into the present. I was reminded of the level of attention to nature that I often see in my son—would it fade and become only an intentional practice instead of instinct?

Years ago, I'd spent some time interviewing an Iñupiat family in Kotzebue, Alaska, about the ways they were coping with climate impacts in their coastal community. One of the many lessons that stuck with me was their mutualistic relationship with nature. *Kamaksriłiq Nunam Inqtananik*, meaning "respect for nature," is one of the core traditional values, *Iñupiat Iḷitqusiat*, that guide the Iñupiat people. *Iḷitqusiat* means "that which makes us who we are."

For 99.9 percent of humanity's existence, people have lived in nature—and in a version of nature that is far more intact and whole than what most people experience today. From an evolutionary perspective, this more urban life is but a blip in time in our species' existence. For me, any effort to visit a tree or a forest is not a journey to see an attraction. It comes from a yearning for the calm I feel in the presence of trees, the sense of coming home. I spent one early morning in Seattle running through Discovery Park, a lush, forested area on the edges of the city and abutting the sea. Wandering through tree tunnels, I felt my mind and body ease. I thought about the scientists who are trying to better understand the connections between human experience in nature—or forests specifically—and well-being and about the doctors who are starting to prescribe "nature contact" to help alleviate our modern mental struggles.[12]

That same day, I went to visit Greg Bratman, the scientist who first introduced me to the idea of using a "dose-response" model for assessing the benefits of nature on various dimensions of human well-being. That was years ago, when we had overlapped at Stanford; he'd studied with Eric Lambin too. Through this approach, subjects are exposed to nature (some sort of regulated dose, like a thirty-minute walk through a forest), and then scientists observe and measure various outcomes (the response, such as intensity of anxiety, rumination, or worry). Traditional ecological knowledge offers a deep understanding of the reciprocal relationship between humans and nature, but Greg quantifies outcomes for people. We were meeting at a picnic table in a courtyard on the University of Washington campus, where Greg is director of the Environment & Well-Being Lab.

"The things I study are already clear to many people," he told me, explaining that he uses tools from Western science to understand effects and relationships that a lot of people and cultures around the world already know to be true. "They also have a use-based relationship baked into them."

Using the dose-response model in one of his earliest studies, he'd experimented with healthy participants taking either a ninety-minute walk in a natural setting (a grassland greenspace with scattered oak trees and shrubs) or in an urban setting (a busy thoroughfare with multiple traffic lanes in each direction). The comparison between groups revealed that participants who went on the nature walk reported lower levels of rumination than those who walked through the urban setting. In the nature walkers, Greg and his colleagues also found reduced neural activity in an area of the brain linked to risk of mental illness.[13]

The fact that he was measuring what people and cultures have known for thousands of years struck me as kind of sad. I took it as another indication that in this modern industrial age, where more people have become disconnected from nature, there is this disheartening need for proof that nature is still essential.

Nevertheless, a lot of relevant research has forged this field of understanding how nature affects human well-being. It doesn't even have to be an experience *in* nature. Take, for example, a seminal study published in *Science* by environmental and architectural psychologist Roger Ulrich in 1984. Ulrich found that simply having a window *view* of nature can accelerate healing and recovery time for hospital patients. Among researchers like Greg, he is still widely recognized for his Stress Reduction Theory, which holds that natural environments promote recovery from stress. The theory suggests that landscapes with views of water or vegetation, with depth and complexity, were most beneficial for survival of our ancient ancestors; they allowed for spotting food sources and predators. These landscapes still help moderate our mental state and the flow of negative thoughts. The natural environment—trees included—also contributes to people's sense of place.[14]

In the courtyard, he told me about a new study underway, led by one of his students. They'd brought study participants to an old-growth forest and filtered the forest air to regulate their exposure to terpenes, which are volatile compounds that give plants their distinctive smells. Common terpenes such as pinene and linalool are hypothesized to reduce anxiety and regulate mood or pain. They were trying to unpack the physiological outcomes of *Shinrin-Yoku*, or *forest bathing*, the term coined for the Japanese tradition of spending time in a forest. The research team was examining self-reported mood, inflammatory biomarkers, and responses in the parasympathetic nervous system, which controls the body's ability to relax and also plays a part in regulating digestion, heart rate, and breathing.[15]

"The participants wear what's called a PAPR helmet that filters the forest air," Greg told me. "It kind of looks like a space helmet with a filter attached to it that attaches to their waist. They're still getting air"—"Oh, good!" I interrupted, and we laughed—"but we're able to control their exposure to the terpenes."

I envisioned a sort of eco-comic strip, with a group of people sprawled out on logs or beneath canopies, trying to soak in the calming effects of a lovely forest bath while wearing space suits like astronauts on the moon.

So, this is what we've come to? I felt captivated by the science, this frontier of understanding of forest benefits, but also perturbed by the need to keep adding reasons for why people need trees. In some ways, experiments like the terpene study have been a response to the forest degradation that continues despite this new movement to restore. He said, "The more we know about all these benefits in a quantifiable way, the more we know about what we're losing when we lose it." Yet, more optimistically, this line of research might also tell us what we're gaining when we gain more forests back.

Later, I tracked down medical doctors who are working at this nexus of public health, ecology, and well-being. Pooja Tandon is a practicing pediatrician, a researcher also affiliated with the University of Washington, and the director of health at the Trust for Public Land. She told me that in the last decade there's been a lot of scholarly interest in "nature contact" and the outcomes for physical and mental health of living near or having experience in nature. With research supporting a positive relationship between nature contact and improved mental state, more physical activity, or even decreased symptoms of ADHD, she and other doctors can prescribe, for example, time in local parks or green spaces as part of a treatment plan. Pediatricians can use a well-child visit to encourage nature contact. A lot of the policy work she's been involved in focuses on facilitating more equitable access; if kids don't have a green space nearby or "nature-rich spaces" at their school, they can't do what even a doctor might recommend.

Studies in this arena commonly assess how experiences in different settings, such as paved versus natural areas or "green spaces," affect mental health. Others go further to determine how variations in fine-scale characteristics, like the amount of tree cover or the forest type, may be associated with different outcomes. There are other efforts underway to leverage and test the mental health benefits of engaging in the act of restoring forests. I found a project in Tasmania, for example, that had restored about 700 hectares of plantation land to native forest. Those involved reported benefits to their physical and mental health. Pooja had another colleague who was starting a pilot program with an after-school program in eastern Washington. Its mission is to teach kids social-emotional learning and resilience through ecosystem restoration. Removing invasive species and planting trees offer the children metaphors for coping with "messy stuff" and "starting afresh."[16]

As I left my meeting with Greg and made my way through the city streets, I dwelled on the expanded list of what we're losing because of climate change, development, and this fast-paced, "modern" form of life. But that same list describes the broader, unrealized potential of forests and the dream for recovery, in relation not only to the imbalances in the atmosphere but also to the ones inside ourselves.

Trees are the load-bearing walls of this house we call home.

I left Seattle by driving beneath arches of fading fall colors that stretched across busy roads. About forty minutes later, I crossed a steel-truss bridge that framed fuming smokestacks in the distance as I reached the outskirts of Tacoma, one of the least-treed cities in the Evergreen State. I felt the heavy weight of inequities born from green and red lines, drawn nearly a century ago. I also felt acutely aware of my own privilege and how it has shaped my perception of what a forest is or should be. Smelling something burning, I pulled over for a moment beside the harbor, fearing that my rental car was smoking. False alarm; the smell was from the industrial complexes across the water, not far from the quainter and more historic old city with red-bricked walkways and Italianate architecture. Mustafi Santiago Ali, executive vice president of the National Wildlife Federation, calls the places where people put everything that nobody wants "sacrifice zones." Such high-pollutant areas, plagued by powerplants, incinerators, and other contaminants, have always been home to communities of color.[17]

The first sawmill in Tacoma began operating in the 1850s. By 1907, there were more than twenty mills on the bay, working twenty-four hours a day, seven days a week. The city became known as the "Lumber Capital of the World." By the end of the twentieth century, the Port of Tacoma had become the main gateway for general trade between Asia and the central, eastern, and northwestern United States. For almost 100 years, a copper smelter operated there as well, spewing arsenic, lead, and other heavy metals across the region. Some reports have shown that air pollution from the smelter settled on the soil surface across more than 1,000 square miles. Contaminants can still be found in soils throughout the Puget Sound basin today. The smokestacks I saw on my drive into town were part of the WestRock paper mill, one of the last vestiges of the logging industry.[18]

"Now here we are, trying to put back a forest," Lowell Wyse told me later that day. Lowell is a literary and environmental scholar and the executive director of the Tacoma Tree Foundation, a relatively new local organization aimed at growing the urban forest. It's targeting another reversal, a local transition from loss to gain.

Driving around Tacoma that morning, I could see some parts were far more desolate than others in terms of trees. The city's foresters want to raise the urban tree cover from an average of 20 percent to 30 percent by 2030, but only 11 percent of the land area is plantable.

I visited a schoolyard where about sixty trees were growing alongside a barrier wall that separates the recess area from the freeway. School kids had helped plant them in 2017, as part of Green Tacoma Day, an annual event. Some of the trees were now reaching up past the wall toward the traffic, which billowed fumes nearby. I watched the kids kick soccer balls and chase each other across asphalt and concrete. The landscape didn't look like what I have known as forest, per se. *I hope those trees are doing all their jobs*, I thought. Then, *I wish there were more.*

That row of trees was a sound barrier and a protector from pollution, but it wasn't a refuge, a forest to walk within, a canopy to climb. There was so much more it could and should be. Perhaps it is easiest to see trees as a fundamental right when they are for children.

"Targeting a 10 percent increase with 11 percent available land basically means filling in everything we have, until you start pulling out pavement to create more space," Lowell told me. "In areas where neighborhoods are completely paved over, there's no way to get that extra 10 percent, or the more that's needed if we're aiming for equity."

Through various tree giveaway and volunteer programs, about 1,500 trees had been planted that year. Mike Carey, the city's Urban Forest Program manager, found they'd need a few hundred thousand big trees or a *million* small trees to reach 30 percent cover.

"We're not going to reach those goals," Lowell told me as we walked around a neighborhood with about 10 percent cover. The sidewalks were cracked in places. Cars drove by blasting music and beats. "But what we're finding is the relationships and the community engagement have huge benefits," he continued. "If you can get to be stewards of a place and care about the physical environment and ask questions, like, 'How can I make my community better?' then we're still really doing something." He makes a good argument for why targets for numbers of trees planted or a percentage cover increase overlook much more that can come from the process of doing this work and from individual trees themselves.

"Trees are symbolic; they really represent something for people. I see this trillion trees movement as an opportunity for buy-in. What is it about trees that make people comfortable saying yes even if they hate environmental policy? We get stories about people who had to cut a tree because it was too close to their house, but then they'll say, 'Sure, I'll take that tree and plant it.' There's something more philosophical there."

The philosophical doesn't always translate to the very practical for long-term growth. Some cities, like Tacoma, will plant, water, and maintain a tree for a few years but then transfer the responsibility to the adjacent property owner. If someone is dealing with stressful life circumstances or other demands on their time and resources, caring for that tree may not be possible. Lowell said that to really move the needle on tree canopy, perception of trees and funding for their maintenance need to shift dramatically. Then people would accept trees as a norm, like they do streetlights and stop signs, and the city would maintain those trees as a standard practice. "This is essential infrastructure that's about our public health and safety too."

When I later pressed him for any examples of resistance, he mentioned only one he had experienced. Some 200 trees had been removed after volunteers from a corporation had planted them. Lowell felt devastated and also hesitant to try the same location again. He'd suspected the removal was an act of vandalism. The representative at the funding corporation had been understanding and said the volunteers still had had a great experience. "It was the only project we did all year that didn't have local involvement," Lowell told me.

That afternoon, Lowell and I walked around the city with a couple of scientists and other staff from the Washington chapter of The Nature Conservancy. They had teamed up with the Tacoma Tree Foundation and Greg's lab at the University of Washington to study changes in surface temperature and human well-being as tree planting in the neighborhood ramped up. That summer, they'd already observed large differences in the heat island effect across the neighborhood.

I asked Lowell if there were any sites in the city that inspired him or gave him hope. He told me about a block a few miles away where the neighbors had come together to plant trees and create a green space. He called it an "unsanctioned community-driven project," meaning it wasn't authorized by the city or municipality. Marked for future development, the land had been laying fallow. The neighbors and adjacent landowners saw an opportunity to create something different.

Lowell said they'd planted Douglas fir and western red cedar. There were big leaf maple trees, beech, and other species too.

"They even planted a few giant sequoias."

That stopped me in my tracks on the city streets. I thought of Calder's seeds, the groves in California, and the many "ambassadors," as Christy had referred to them, that people have scattered across the planet.

"Will you take me there?"

He agreed, and we headed back toward our cars for the short drive.

"I'll follow you to the giant sequoias," I told him.

Lowell grinned. "Well, they're not giant yet."

We found our way to another residential neighborhood, which, Lowell explained, had been historically underserved and redlined. Across the street from where we parked, a ball field was fenced off to the public, accessible only for specific team events and programs.[19]

"It's literally locked," Lowell said. "Heartbreaking, really. I've watched kids playing catch on the asphalt in the parking lot because they can't get onto the field."

We walked past an empty lot, and Lowell pointed out a little area that had been "de-paved." Volunteers had busted up the asphalt to make room for shrubs. Lowell and I meandered through a fenced alleyway and emerged onto a striking green field with saplings and young trees scattered throughout. The neighbors and adjacent landowners had been caring for the trees and mowing the field for a couple years now. Hand in hand, a couple was making a lap of the urban forest together. My pace slowed down as I relaxed into a stroll.

Assuming the bird's-eye or satellite perspective, I could visualize that land as only a couple pixels on the planet. But walking there with Lowell, I felt what it meant for many lives connected to it.

Compared to the intact forests remaining or any large landscape planted to maximize carbon sequestration, urban forests represent a relatively small but important corner of the solution space for the climate crisis. Yet the 500 million trees it would take to achieve tree equity in US urban areas alone could also absorb nine megatonnes of carbon annually; that's equivalent to taking more than seven million cars off the road every year. So, the carbon piece is there, but it's just not often first and foremost when considering all the other benefits that trees provide in the urban context.[20]

"They're specialty trees," said many people who spoke at the conference in Seattle and others I interviewed. They were referencing all the other benefits that can be quantified in the urban communities, from the cooling effects to reduced energy consumption. Some programs might cover the costs of getting trees in the ground, but it's the people in place who often determine their future. Those people—living in an urban forest or near a big block of intact forest too—need to see and feel and believe in the benefits of the trees to ensure their survival.

9

TREE TIME

In June 2021, I came across an article in *Forbes* magazine showing that restoration, typically an underfunded, underserved, and undervalued endeavor, had entered another realm. The article was about Yishan Wong, a Silicon Valley veteran, and a start-up he'd formed not long after the Global Potential study exploded in the media. The company, Terraformation, wasn't focused on some new app or internet platform, as I might have expected from the former CEO of Reddit. *Forbes* called it "a startup which aims to help companies and countries meet zero carbon goals of the future through rapid reforestation." At the time, the Hawai'i-based company had just received $30 million in funding that would support its mission to help restore native ecosystems around the world, using technology such as solar-powered desalination and other tools it would later develop. It wasn't a philanthropic effort or an organization funded primarily by grants or government money; it was a for-profit company entering the global reforestation movement. To me, this signaled a new awakening to the value of trees and all that forests deliver. But it was also some form of *re*-awakening of what people have known for millennia. In this industrial age, far too many of us have lost track of how life depends on functioning natural ecosystems—those load-bearing walls of this house we call home.[1]

As the story goes, and my interviews with Yee Lee, Terraformation's vice president of growth, and other staff later confirmed, Yishan, his friends, and colleagues had become increasingly concerned about climate change and doubtful that governments would take effective action for long-term problem solving. Firsthand experience of rampant wildfires had intensified their children's worries too. He'd run some back-of-the-envelope calculations on various strategies for sequestering carbon and other greenhouse gases. He tried estimating how many trees it might take to offset global emissions.

"We came to this idea of focusing on forestry after researching many different solutions," Yee told me of their collaboration, before diving into a slew of examples. "We talked with the folks who were crushing up limestone, sprinkling it on equatorial beaches to change ocean acidity and absorb more carbon dioxide from the atmosphere. That was kind of interesting! But will it scale? How do you control it in case it starts to produce side effects that you don't like? There weren't good answers to any of these questions." He said the same was true for a lot of other ideas, then quickly ran through a list.

"There's a ton of people working on many different novel solutions, bioengineering trees, direct carbon capture, of course, and aerosol sprays to change atmospheric albedo." He was referring to the idea of spraying mass quantities of tiny particles into the stratosphere, an upper layer of the atmosphere, to cool the planet by reflecting sunlight back into space. "There are so many of these very science-fictiony-sounding solutions, and none of them meet the real basic criteria. Like, is there an actual economic incentive for anybody around the planet to do this, or is it just a straight cost or an experimental risk?" He mentioned the very high cost of direct carbon capture, which was making its uptake by individuals prohibitively expensive at the time. Such technological approaches that offer long-term storage of carbon dioxide in various forms could take decades to scale. Fans that extract carbon dioxide from the atmosphere at a given location also require energy inputs to work. He noted, "There's no economic incentive for somebody to set up a giant fan in their backyard!"

Yishan told the *Forbes* reporter that because he was a computer scientist and not an environmental scientist, he didn't trust his math on the trees. He'd sat on the idea for a while, until the Global Potential study confirmed the numbers.

Yishan's very simple math, which he shared on August 17, 2019, on *Medium*, the online publishing platform, went like this:

First, the basic unit of CO_2 is a "ton." A billion tons is a gigaton. Each year, the world emits about 45 billion tons, or 45 gigatons of CO_2.

We know that forests absorb CO_2. How much? An acre of forest absorbs about 15 tons of CO_2 in a year. Other tree species absorb somewhat more or less, some of them a lot more.

This means we need 3 billion acres of forest to offset our entire worldwide CO_2 emissions each year.[2]

That 3 billion acres is about 1.2 billion hectares, an area larger than the United States, and—when accounting for an average of 100 trees per hectare—1.2 trillion trees. His calculations yielded an estimate of *need* that was higher than the estimate of *potential* that Tom Crowther, Jean-François Bastin, and their colleagues had revealed: "room for an extra 0.9 billion hectares of canopy cover." Nevertheless, Yishan's problem and solution were yet another affirmation for people to target a trillion trees.

"Yishan is your nerd's nerd," Yee declared to me with a tone of pride. Ask him to calculate something, and he starts with the assumptions and fundamental principles and goes from there. It ended up that there were a lot of parties triangulating around similar numbers for trees and their carbon-sequestering potential.

"Forests turn out to be a perfect engineering solution, from a geoengineering perspective," Yee said. "This was literally a slide in our venture pitch deck when we started Terraformation. It was basically saying, 'If we came in here and said we invented a self-replicating, solar-powered, carbon dioxide absorbing machine that poops out oxygen and sometimes food and is adapted for almost every biome on the planet, you would be throwing term sheets at us right now!'"

"Hey, it's already here! We know what it is!" I interjected into the ta-da moment.

Yee later showed me the slide explaining that this invention ran on sunlight and water and produced no pollution. But the "technology" already exists, he noted: "It's called a tree."

"The fact that this is a biological entity," Yee told me, "that we've known about for all of human history, and it is the product of billions of years of evolution—well, all that just means this thing is just suffering from a marketing problem. The solution is right in front of us."

The *Forbes* article and others that ran about Terraformation showed pictures of solar panels resting on a drought-stricken landscape with an ocean view in Hawai'i. They'd built an off-grid, solar-powered desalination system that filters brackish water into thousands of gallons of freshwater each day. Given the site conditions and proximity to the ocean, the groundwater is salty before treatment. An irrigation system then distributes the freshwater to the thousands of trees planted at the company's first restoration site.

"We picked a hard engineering challenge to start," Yee explained. "But we thought, 'If we can make this work here, then we can make this work anywhere.'"

They'd learn by doing and correcting course as needed, working fast and hard, as any start-up does.

I could see Yee's sound logic and appreciated his optimistic take on forest regeneration and restoration as essential in aggressive climate action. But this framing of "THE ONE SOLUTION" had already gotten dangerous around the Global Potential study, given the misinterpretation that more trees might alleviate the need for emissions reductions. A number of "yes, but" thoughts stirred in me: Yes, but it's not *the* solution. It's one solution, yes, but it's not so simple. Yes, but we can't trade trees in one place for emissions in another and call that the entire global fix. Yes, but everyone interested in planting trees needs to be targeting more than just the carbon. Yes, but is some of the land that could become forest being used for something else?

About a year and a half after I first heard about Terraformation, I went to Hawai'i Island, the "Big Island," to visit their headquarters and that first restoration site. Putting the company aside for a moment, Hawai'i itself is a fascinating and complicated place to pursue restoration or any vegetation planting at all. The islands are known to some as the extinction capital of the world. Some 400 species are listed as threatened or endangered (that's about one-quarter of all species listed in the entire United States), and about 90 percent of the native flora is endemic—not found anywhere else in the world.[3]

Except for the hoary bat, there are no native land mammals. But spend a day or two exploring the Big Island, for example, and you're sure to come across a wild pig, mongoose, or seemingly more ordinary chicken or cow. Animals, like many plants, came with people who arrived in the remote archipelago. Some species arrived with the Polynesians over 1,500 years ago; others came after the first colonial contact by Captain James Cook in 1778. There were intentional introductions by the Polynesians, like sugarcane, taro, and other food crops. There were also unintentional ones, like the rats that arrived with Cook's ships and others that followed. The mongooses found in Hawai'i today were introduced in the late 1800s with the hope that they would reduce the rat population in the sugarcane fields. Rats are nocturnal, and mongooses are diurnal; so the predation plan failed. They both run rampant today. Strawberry guava from Brazil, for another example, spreads fast throughout forests and prevents other plant growth. It was brought to Hawai'i in the early 1800s for its fruit and attractiveness, but since then it has overrun many native forests.[4]

Without even considering what species might be best suited to survive future climate conditions, deciding what to plant in Hawai'i has a lot of layers. For example, are we trying to go back to what was there before any human contact at all or just before European contact? If we plant what *was* there, how will that interact with what *is* there now?

During my visit, I stayed in a yurt on a coffee farm owned by the family of a friend from Santa Cruz, California. South past Kona, the road narrowed as I made my way to their property. The turquoise ocean was to my right; dense forest lay inland to my left. *Tree tunnels,* I wrote in my journal that first night. *Enormous palmate leaves, swooping canopies, vines dangling. What is invasive? What is native?* I fell asleep celebrating the orchestra of frogs croaking, then learned days later that they, too, were invasive and loathed by many locals. *Woke up at first light with the birds, surrounded by far more species than I can identify. THIS is a forest and a farm.*

It felt more like a forest than a manicured plantation to me, but I also knew it was a battleground for plants that are present versus plants that *should* be present, with a long history of land use, values, chance, and evolution dictating the tensions.

A couple of mornings later, I was riding in the car with Keali'i Thoene, a Native Hawaiian and the community programs manager for Terraformation in Hawai'i. We were heading out to the Pacific Flight restoration site, where the desalination plant chugged day after day and the trees were growing. Keali'i and I had met earlier that morning at the headquarters in Kona; he was wearing a company T-shirt with "Restore the Planet" on the back. Keali'i had the calm demeanor of a surfer, and a few mentions of waves later confirmed my impressions. Within the first few minutes of meeting, he gave me a brief overview of the Hawaiian language and its thirteen letters and of *mo'olelo*, the stories and legends with hidden meaning, *kaona*. They pass from one generation to the next, creating a cultural fabric of Hawai'i. Becky Hart, a self-proclaimed "nature nerd" with a degree in natural resource ecology, sat in the backseat. She was working as the nursery assistant, collecting and storing wild seed and cultivating saplings for Terraformation.[5]

"Why is it called Pacific Flight?" I asked Keali'i as we drove northeast together from Kona. The distant ocean glowed turquoise. The surrounding landscape turned to dark lava fields. Crushed and jagged rock lined the edges of the road.

"That was the first question I asked as community programs manager!" he replied. Yishan had picked that name. It was arbitrary, Keali'i explained, and Yishan hadn't been too attached to it. "I was like, 'Great! We've got a starting point because the place where we are already has a name. It's Kaupalaoa."

I attempted to repeat the beautiful Hawaiian name, my effort hardly resembling the word. We practiced until I got it right.

"So, we decided in the future when we name projects, you can call it anything you want. But you need to include the traditional place-name. So, it is Pacific Flight at Kaupalaoa. We include the place-names so they don't die. The restoration of human cultures is the same as the restoration of our natural ecosystems."

Keali'i said Terraformation's efforts in other countries had their own dynamics and complexities, but he also saw Hawai'i as an experimental model for working out the kinks in native restoration.

"We're trying to be a lighthouse, a shining example—this place where we can demonstrate what restoration looks like. We have a rich history of Indigenous interaction with nature, which is what a lot of other places where forests have been cleared look like too. We have a government structure that can be slow and tedious and bureaucratic, which is also what a lot of other communities experience. We are hoping that solutions can come from here and be tested here too."

We drove through the little town of Waimea and passed fields for cattle grazing.

"This area that we are in right here is called *keala'ōhi'a*, the pathway of *ōhi'a* trees."* With their deep island history and bright red or yellow flowers that look a bit like spikey sea urchins clinging to branches, *ōhi'a lehua* are one of Hawai'i's most iconic trees. They are pioneers, among the first to colonize the harsh lava rock that forms the islands. I scanned the grassy horizon looking for the *ōhi'a*. There were hardly any trees at all, just wide-open pasture.

"This spot right here; that's the Waimea Middle School. But I don't see a single *ōhi'a* tree!" Keali'i exclaimed. "This is another reason for preserving place-names. You might ask, 'What was here before?' Well, it wasn't recorded; it was all in oral histories. Places' names have data locked up in them."

The absence of *ōhi'a* trees and the ecological history captured in a name made me recall the distinct perceptions of "normal" that my son Calder and I have. The ecological baselines that we know are not the same because we have come into being decades apart from one another while the world changed rapidly. I shared this with Keali'i, and he listened intently to my struggle of knowing about the dramatic environmental change that has occurred in just two generations.

* The Hawaiian language is spoken, not written. As Keali'i later pointed out, *ala* and *'ala* sound almost the same, yet have different meanings: *'ala* can mean "fragrant," "perfumed," "esteemed," or even "chiefly"; *ala* can mean "pathway." So, he could say *alaohia* in reference to this powerful place in Waimea, but the interpretation is up to the listener. Fragrant, regal, or a pathway lined with fragrant, regal *ōhi'a* trees. "It is up to us to walk that path and determine how these trees speak to us on any given day," Keali'i clarified.

My instinct has been to spare my son the grief of understanding the extent of loss and degradation and to hold out as long as possible on that. But if everyone turns a blind eye to the changes that have occurred, I also know we'll lose the collective knowledge of a world where people and nature worked together in functioning ecosystems when we weren't overwhelmed by this doomsday possibility of Earth becoming uninhabitable. If I fully accept the current state—Calder's baseline—as my own true normal, where is the motivation to strive for anything different?[6]

When Keali'i had gone searching for the meaning of *Kaupalaoa*, he found it linked to a sperm whale that was born and lived in the sea. From the history, it was guarded by the *'aoa*—sandalwood—in the forest. A name could preserve the history of the land. A name could signpost the species that had been present before and might flourish there again.[7]

He later wrote to me with a follow-up note about our discussion of names in the car.

> I wanted to leave you with one more piece of data tied to the Hawaiian language:
>
> *Hahai nō ka ua i ka ululā'au*—the rain follows the forest.
> This is one of thousands of *'ōlelo no'eau*. You can think of those as, like, proverbs, that have data and lessons, and things to keep you safe, things to keep your community healthy and thriving for many generations. While we did not use words like *evapotranspiration* (though there is a Hawaiian word for that), we did understand how these natural ecological processes are connected, because they are us. It is not a metaphor or a simile.[8]

In tropical and subtropical forests, local climate benefits occur during the dry season, in regions with low rainfall, and during heat waves. Water that trees transpire provides a cooling service. The big blocks in the tropics provide water vapor that supports cloud formation, which also has far-reaching effects on temperatures and precipitation.

Keali'i said all things in the natural world are our relatives.

"So, when our forest relative is gone, our rain friend is gone too because *Hahai nō ka ua i ka ululā'au*—the rain follows the forest."

The sun was shining on Kaupalaoa when we arrived via a winding road. It was warm and dry, and the hardpan soils, compacted over years, made a crunching sound as we walked across the property. The entire side of Kohala Mountain, where the restoration site comprises a sliver of land, had been acacia and sandalwood forest long ago. The sandalwood was the first to go in the early 1800s; it was exported to China. Acacia went next, and eventually the land had been converted to pasture. But without enough grass, it was only marginal for that purpose. The land looked somewhat barren and windswept from a distance. Up close, I could see small shrubs and skinny saplings—wiliwili (*Erythrina sandwicensis*), a'ali'i (*Dodonae viscosa*), and koai'a (*Acacia koai'a*).[9]

"The planting process here is pretty labor intensive," said Bryn Lawrence, the site operations manager, as we began talking over the details. He said they'd planted about 9,000 trees and shrubs, starting in 2020, and about 7,000 individuals were still living. I had seen online sources and news articles that reported 7,200 trees planted. They backfilled the dead ones with new trees when replacements were ready in the nursery and the timing was right for planting. Bryn seemed genuinely pleased to have visitors. He'd been living year-round on the site and was responsible for keeping the desalination, or "desal," and irrigation functioning together around the clock.

"We're auguring a hole with a six-inch drill bit. We make a six-inch-wide, deep hole for each plant that gets new soil and fertilizer and water that same day. Then it's irrigated moving forward." Bryn's background is in biology and toxicology, but he said that he was now doing a lot more engineering and monitoring.

"It's really important to check everything frequently, so that when there's a problem, you discover it right away." He briefly detailed how he monitors water use, water production, and the intricate workings of the solar-powered desal system, along with regularly checking for irrigation leaks.

The plan was not to irrigate the trees forever. Kaupalaoa is located on the leeward side of the island, so even the restored forest, if successful, would be prepared for the relatively dry conditions. But *Hahai nō ka ua i ka ululā'au*—the rain follows the forest. If more restoration occurred on sites across the island, clouds might eventually hover and bring more mist on a regular basis. Yet Hawai'i is also generally getting drier with climate change. I wondered about the timeline for tree growth and whether our rain friend would come back with the trees before the predicted warmer and drier conditions stressed the local dynamics further.[10]

The desal room was a retrofitted garage, built next to the brackish water well on site. The pumps and power systems were so loud inside that we agreed in advance to hold any questions until we'd made our way to the electrical room, which stored and distributed the solar-generated electricity. The space was meticulously clean, and the filtration system reminded me of some sort of Dr. Seuss invention: pipes ran in numerous directions and connected to orange hubs, tanks, gauges, and valves with red handles. I couldn't discern what did what, but it was still captivating. Keali'i went to a cabinet, took out a glass, then approached a small valve on one side. With a proud grin, he turned a handle, which opened the flow, and poured me a glass of freshwater.

"It's flavorless," I said, amazed and a bit confused, as I'd never thought of water as having flavor. It was weird. "It's like wet air," I added. Everyone laughed.

The desalination process removes all minerals, so it was probably the purest H_2O that I'd ever sipped. To bring the pH back to normal, they add calcite back into the water. Outside, they also put a small amount of bleach in the 50,000-gallon holding tanks to limit any bacterial growth. The bleach is heavily diluted.

Inside the electrical room, I stood before a wall of giant white blocks. They were mounted in a grid on the far side of the room. "Blue Ion" in bright blue lettering repeated across the encasings up top, and a screen reported the amount of energy stored inside the batteries at 98 percent.

I found myself wondering about the cost of it all. Yee had been frank that desal for forest restoration at scale wasn't very feasible under their current business model. I knew Terraformation was attracting corporate funders who were interested in buying carbon credits to offset their own emissions. The companies would effectively fund forest restoration efforts to be able to claim the amount of carbon that the trees sequester over time against their own emissions.

"Are there other projects you're pursuing that are using desalination?" I'd asked Yee before my trip to Hawai'i.

"We have not had to employ reverse osmosis in any other site," he'd said. "We've come across sites that have requested our assistance with that, but you need to pencil out the economics of it. Maybe if you can use the water for other commercial purposes or home water supply or drinking water, then it might work. But then you're not just doing water for forestry, you're basically running a water company now. So, you've got to think about how much you want to invest and how dedicated you are to running a water company. It's a costly source of natural resources to help grow a forest."

"But it is an example of innovating in a place to bring back a forest," I'd commented. "Does it seem like something scalable?" He'd said it was scalable if someone had enough money and time. When Yishan Wong had published that post on *Medium* with the simple math for three billion acres of trees, he'd also written about possible approaches for reforesting deserts. There had been interest from people in the Middle East who wanted to regreen the desert, but Terraformation wasn't pursuing the endeavor.[11]

"Our bread and butter is trying to find the places that are really amenable to forest restoration—so, tropical forest restoration, mangrove restoration," Yee had explained. "We focus on areas like Central America, West Coast of Africa, the Southeast Asia Corridor, for example. These are regions where we can recover lands from commercial use like timber plantations or overgrazing and turn them back into native forest." Part of that process, he'd said, was developing forest-based business models that sustain the local activities and also create new livelihoods.

Yishan had installed the desalination system for freshwater when he'd originally purchased the property and intended to build several homes. That was in 2017, before the estimate by Jean-François Bastin and his colleagues on the potential of trees confirmed Yishan's simple math and triggered a quick turnaround in his thinking. Yishan then abandoned the initial plan, turned the property into a forest restoration site, and, in 2020, founded the company with a global restoration mission.

Bryn, Keali'i, Becky, and I walked the property together. Near the gravel driveway, there was a dark blue shipping container alongside some other sort of filtration system with tanks and connecting hoses. It was the aftermath of another experiment: a company called Heimdal had been extracting minerals and acids from the brine biproduct on site. It had been exploring ways of rebalancing ocean acidity by transforming the dissolved carbon dioxide into calcium and magnesium carbonates that would then fall to the ocean floor. It was a clever idea for using the extra salty brine that was a leftover from the desalination process.

"It didn't work?" I asked.

"It worked. It worked," Bryn said. The issue, apparently, was finding a way to operate at scale and make that profitable.

"Scaling!" Becky grumbled, seeming slightly exasperated. "That's always the problem."

We kept walking, and she identified species from the wide selection of native ones they'd planted. We found our way to a little spindly *ōhi'a* tree. I thought about this notion of Hawai'i as a "lighthouse" for restoration that Keali'i had mentioned. I thought about the time it would take for this new story of Kaupalaoa and its potential forest to unfold.

What does this place illuminate?

It all seemed quite extreme: the drought-stricken land, the compacted and nutrient-depleted soils, the brackish water turned to freshwater through the engineered interaction between technology and the sun, the fate of the native trees amidst everything nonnative established on the island, the year-round caretaker dedicated to the trees, and the Silicon Valley veterans turned tree huggers endeavoring to create a business model that incentivizes lasting forests, increases forest carbon capture at scale, and sustains biodiversity through the habitat created.

It also brought to light the many challenges of planting trees in hope of creating future forests anywhere else. What to plant? Where to plant? When to plant? How to support the trees' establishment? How to monitor and care for them? The investment needed over time—in terms of not just money but personal commitments from near and far. The people living in and around where trees and forests could be must also want the trees. They must see and feel and believe in the benefit from the trees and work to help them grow. Some tangible value or values must drive it all and be shared by many people, in one form or another.

As we departed Kaupalaoa, Keali'i showed me a slide from his files on his phone. It was a grid of four core values central to Terraformation's mission. One focused on our ancestors, acknowledging that our choices today will shape the world for generations: "Let's make our descendants proud." That one hit home for me as a mother. I cannot correct for all the degradation that has transpired from decades of increasing human activity wrought upon the Earth under capitalism. Yet, at some point, I'd like to be able to tell my son that I tried to make things better, and I tried to play one tiny part in the many solutions needed to redirect the future course.

Another emphasized understanding. Great solutions come from a deep understanding of a problem, it noted. We must "gather many viewpoints and consider the thoughts of people who may not be in the room." Biodiversity was the third core value; the richness of species is what makes an ecosystem resilient. I was on board with those values too; the notion that "biodiversity is essential" was nothing profound to me as an ecologist. But the last one implicitly captured a broader awakening to the colliding crises of persistent forest loss and climate change, and I felt inspired by that.

"Start the impossible," it stated. "Some solutions present themselves only after starting down the path. Don't be afraid to begin because you can't see the end."

Not trying to problem-solve may be another kind of sin, even if there are persistent debates surrounding how best to harness the potential of trees and uncertainties surrounding other mitigation prospects as well. Given my experience working for a few environmental NGOs over the years, I have grown accustomed to the challenges of tackling urgent problems with limited budgets and, at times, with frustratingly slow operations. Indeed, there was something both refreshing and intriguing about the new corporate interest in the work that has typically fallen to ecologists, foresters, and environmentalists in the recent past and to Indigenous people for millennia.

"People love to plant trees, but they don't want to come back to pull the weeds," Will Weaver told me. We were sitting on a couch in one of the other yurts on the farm where I was staying. Wili, as my hosts called him, was working for a local restoration company, Forest Solutions. He had come for dinner with friends, but I couldn't resist firing off questions, given his nearly twenty years of working in conservation and restoration in Hawai'i.

Wili had quite a portfolio of projects under his belt. He'd helped to plant native koa trees for a guitar company, collected seed of rare plant species by helicopter to bring them back from the brink, and worked with private and public landowners to protect and restore forests in the Ko'olau Mountains on O'ahu, to name a few. He was setting up training in forestry to build local capacity for the increasing demand and exploring ideas to start an orchard at his parents' property to produce more seed.

"Anything for carbon?" I asked. "As in, planting trees with sequestration as the primary goal?" I was keen to see if the idea for mass plantings—largely motivated by carbon interests—had taken hold in other parts of Hawai'i.

He grimaced a bit and mentioned one project on O'ahu, led in part by a professor from Colombia at the University of Hawai'i, Camilo Mora.

"Mora cleared the land Amazon style, and then we worked with volunteers to plant thousands of native trees." He hadn't burned the land, but invasives had grown so thick that they had to clear the property with chainsaws and a mulcher before planting. The spacing of the trees deviated from what other local restoration experts would have done. The plan had targeted a forest with big trees, overlooking the fact that planting closer together would create a more closed canopy with smaller trees that could lock out light and help restore the soil. Invasives also returned without adequate follow-up maintenance.

"He was trying to grow those trees far too quickly and missed the mark," Wili told me. Mora had seemed too focused on the potential for more carbon sequestration, promoting calculations based on a lifetime—decades of tree growth that hadn't happened yet. Big trees have the appeal of big biomass in the larger trunks (generally translating to more carbon stored), but planting trees far apart in anticipation of that outcome doesn't always work. The survival rate of the plantings was low for several reasons; about 50 percent didn't make it from the get-go.

"There's a forest there now, but that's because people came out of the woodwork to help."

They replanted trees. They watered. They removed persistent invasives. Planting was only one step in the process; so much more hard work came afterward.

Wili and other local ecologists whom I later interviewed said the outcomes of the effort were sugarcoated. Mora declared a mission to plant a million trees per year. He coauthored a viewpoint for *Nature Reviews Earth & Environment*, a highly respected journal, asserting that fixing climate change would be much simpler if we all started planting trees now. The article stated, "There are over 2 billion acres of land [~800 million hectares] that could be planted without conflicts with urban development or farming"; the scale of the endeavor appears "daunting" but "equates to just ~130 trees per person, or a 10-hour job for an average individual!" In one online profile, Mora described a set of weighing scales that have become off balance. Carbon emissions weigh down one side; the other side is practically weightless due to the lack of trees. Planting trees, he said, could balance the scale.[12]

I could see that in his planting endeavors and broader communication, Camilo Mora had fallen into the common trap of reporting the number of trees planted, as if that were the end of the story for success. I contacted Mora for his take. Mora said he'd expected the challenges would be finding people to help and getting access to land. There were a lot of eager volunteers. Private landowners were keen to support more native forest. The real challenges were the technical aspects of care for transplanting the saplings and facilitating their survival. "I always thought publishing in *Science* or *Nature* was the hardest thing," he told me, "but nothing compares to how hard it is to plant trees." He said this work wasn't even his job; he'd gotten interested in reforestation to take personal action, amidst the devastating results his climate- and biodiversity-related research had been showing for years.

When I asked him about the 50 percent survival rate, he corrected me: that was only in the initial shock period after planting. About a month later, only 5 percent had survived. I was stunned. People replanted, and the next wave did fare better. Mora didn't seem fazed by what others would call a failure.

"A species like koa produces about 100,000 seeds in its lifetime, and that's to ensure that at least one will make it over time," he explained. "So, even if we get 5 percent, that's 5,000 out of 100,000. Basically, we're doing 5,000 times better than nature would do!" I didn't see it that way, but I could get on board with the ideas of partial success—some form of better—and learning by doing. Wili later told me that, in retrospect, Mora's efforts highlighted the good intention of planting as well as the importance of developing and instilling stewardship into the process.

"The fact that something can be done doesn't mean the tools we have are going to work; we might need to innovate," Mora said. "We had planes, but could we use them to get to the moon?" He had been frustrated by all the harsh criticism. "I don't get it. Make it constructive. Come help us!"

The night before I left the yurt, I lay in bed listening to those invasive frogs again. I was thinking about the urgency of the climate crisis in relation to "tree time," a concept that had arisen briefly in one of my many conversations with urban foresters. I had been speaking with a stewardship coordinator in Chicago before my travels to Seattle and Tacoma.

Her point was that relative to the pressing need for emissions reductions today, tree time is very slow. It is not a day of planting or even the first few years of growth. Working on tree time is not an immediate fix but an investment toward reclaiming balance over decades to come. I recalled a response that I'd read in 2020 to the Global Potential study. "Bastin et al. clearly show that we have a narrow window of time in which to restore global tree cover," it had argued. "We need to act quickly, intelligently, holistically, and globally."[13]

The stewardship coordinator had suggested that finding ways to plant a diversity of species and more equitably distributed trees couldn't just happen at such a grand scale all at once. Getting more people involved would support a movement over time. But a fast-paced mission with parachute planting, dropping into communities to plant trees for what someone from the outside presumed to be a greater good, wouldn't be sustainable.

I thought of Becky Hart at Kaupalaoa, who had declared, "Scaling! That's always the problem." How can all these new players with good intentions and great ambition do it?

All this fuss about harnessing the potential of trees may be more about finding the most effective way to mobilize people—across communities and cultures, companies, and governments—to work together as never before. If enough people are doing the work with a local focus, the coordinator in Chicago had told me, then they form a global village.

I still wondered, *In ten or twenty years' time, or even in tree time, will we get what everyone involved is expecting?*

10
FAIRY TALES

I began curating Calder's first bookshelf when he was a baby, and one of the "tree books" I bought was titled *Planting the Trees of Kenya: The Story of Wangari Maathai.*[1] The book is a story about mutual care: about leadership and problem solving, responding to degradation and tree loss, and motivating women, children, and even soldiers to collect seeds and plant trees. The people in Maathai's community no longer had clean drinking water because the land use had contaminated the watersheds. Nor did they have firewood for cooking. People had cut down their fig trees. They'd forgotten to care for the land that was feeding them, to renew and respect. Maathai saw the potential of communities to re-create what was lost, to foster a better tomorrow by planting trees.

Calder and I spent many wintry evenings reading that book together when he was just sixteen or seventeen months old. He liked to pause on the title page, taking in the drawing of a young girl, dressed in a bright yellow shirt and blue skirt. She holds a sprig of leaves in one hand. As I read aloud, he studied the watercolor illustrations, pointing to colorful pages of rusty red soils and bare, rolling hills, of rivers flowing tan and tainted by sediment.

"When the soil is exposed," Maathai says, "it is crying out for help, it is naked and needs to be clothed in its dress. That is the nature of the land. It needs color, it needs its cloth of green."

I loved the children's fairy tale version of her story and its motivating message of implementing this "simple and big idea." But every time

I read that book, I found myself thinking about that perceived simplicity of planting trees to restore watersheds or reshape local climate, to protect other species, to reduce pollution or erosion, or to suck carbon out of the atmosphere. It all belies a tremendous amount of complexity.

Maathai started an organization called the Green Belt Movement, which still focuses today on restoring degraded watersheds in Kenyan communities. Since the work began in 1977, tens of thousands of women have been trained in forestry, food processing, beekeeping, and other trades to bolster income while also protecting and restoring lands in Kenya. Even in the children's version of the story, planting was just a fraction of that work. There were challenges: An early nursery in her backyard failed, and almost all the seedlings died. Water was limited. Women dug deep holes to access groundwater and carried heavy buckets full of water on their heads.

The act of putting a seed or a sapling in the ground may be relatively uncomplicated, but I don't think anyone should be told to do that or have it done for them without their consent. If planting trees or protecting existing forests or restoring degraded lands means taking any of that land out of productive use—as with agriculture in rural lands or even cement-covered lots in cities—the choice for trees needs to be economically viable. This is the great conundrum about how to balance the approaches to land use that offer the most benefits over generations, including those that traditionally have not been valued by our economic system, and those that deliver the most short-term gains.

Even a young reader can see from Maathai's story that any reversal, big or small, is a movement itself, changing people's relationship with the land and each other. So, I've felt an uncomfortable tension between the ambitious global goals for restoration and the local effort, leadership, and support required for success. The fairy tale for me was this idea that every effort might look like Maathai's, coming from a desire to better lives by repairing nature and sustained by local people over decades.

The schedule for any annual UN climate change conference, or conference of the parties (COP), extends over several days and is packed with high-level meetings, technical sessions, side events, and more. The World

Leaders Summit is often the most anticipated event; it's the time when heads of state and government unveil major climate commitments to kick off discussions and negotiations. The summit is held in a massive convention center. There's a main stage with a formal podium and row upon row of executive chairs behind desks for leaders from around the world. Attendees wear headphones to listen to the speakers in their chosen language. They've flown from all over the world (yes, consuming fossil fuels) to declare what they are doing or to discuss what they might do to combat climate change.

COP27, which took place in Egypt in November 2022, had some 35,000 registered attendees. As in other years, they were not only government representatives but also individuals from "observer organizations," such as nongovernmental organizations and intergovernmental organizations. I wasn't there, but I followed the press releases and media headlines. I later spent hours combing through online clips to find a recording of the launch of the Forests and Climate Leadership Partnership (FCLP), a commitment to halt and reverse forest loss and degradation and deliver on the Glasgow Leaders' Declaration on Forests and Land Use that 141 countries had signed the prior year. Those countries represented 90 percent of the forest cover in the world; they committed to working "collectively to halt and reverse forest loss and land degradation by 2030 while delivering sustainable development and promoting an inclusive rural transformation."[2]

The FCLP summit began with a moving video of a stunning forest illuminated by a ray of light coming from a break in the canopy. Birds chirped in the background. A dramatic voice contrasted tree cover from past to present: "Ten thousand years ago Earth was home to an estimated six trillion trees. By the end of the last century, that number had halved." The images changed to an aerial of clearcut forest, trees splayed on the Earth's surface like toothpicks scattered on the floor.[3]

As one COP attendee told me—and the amount of time it took me to find the recording confirmed—the announcement of the FCLP was "an announcement as part of a bunch of announcements in a day filled with announcements." But it was the focal point that year for anyone working to improve the state of the world's forests or concerned about all the life connected to them.

Press releases and posts on various sites called the FCLP a "high-level partnership" dedicated to delivery and accountability of the Declaration on Forests and Land Use. The United States and Ghana would cochair the effort. Public and private donors pledged to mobilize another $4.5 billion in effort to reach $12 billion toward forest-related programs over a five-year period.[4]

Jad Daley, CEO of American Forests, was there in the room in Sharm El-Sheikh when the partnership rolled out. Jad and I used to work together as advisors to a fund that supports climate-adaptation efforts in the United States. He's always been very insightful and open with me. At the launch, the presidents of Colombia, the Democratic Republic of Congo (DRC), and Ghana and the prime ministers of the United Kingdom, Norway, and the DRC, among others, spoke.[5]

"Just being really candid about it, it didn't feel as climactic as I think it should have," Jad told me over a video call. He was in Washington, DC, at the time. "I don't quite know why or what should have been different, but I think it felt a little bit like this laundry list of commitments, without enough information on how to do the work." He found himself wondering what they all might mean when added together.

"Suffice it to say, there's a lot of pent-up commitments and pledges of different kinds that have been made—more than most people realize."

The Land Gap Report, published right before the conference, assessed the land needed to meet the projected amount of carbon removed from the national commitments and pledges made already. That area was almost 1.2 billion hectares—consistent with what Yishan Wong had promoted as a solution on *Medium* and more than what Jean-François Bastin and his colleagues had reported for the potential in *Science*. It represents a lot of ambition in the hopes of doing something good.[6]

Jad said, "I keep coming back to this: Pledges are fantastic. Commitments are great. But the implementation of them and the fact that they actually happen, and then happen in the right way is really where the action is and needs to be at this moment in time."

The executive summary of the Land Gap Report stated that the amount of land needed to meet the pledges was more than the amount of current global cropland. The authors, twenty researchers from around the world, affirmed that countries' climate pledges "rely on unrealistic amounts of land-based carbon removal." The word *removal* refers to solutions that take carbon out of the atmosphere and lock it away, as opposed to emissions *reductions*, such as cuts achieved through more fuel-efficient cars. About half the estimated area needed for the pledges would require land-use change, with the potential to displace food production and affect livelihoods for smallholder farmers. The other half (slightly less) would restore degraded ecosystems.

I recalled a formula on land scarcity for conservation from graduate school that had been presented in an article by Eric Lambin, my former advisor who won the 2019 Blue Planet Prize, and Patrick Meyfroidt, another Belgian scientist who studies land use and forest transitions:

> Land for nature = Total land area – (Agricultural area + Settlements)[7]

It was what scientists call a conceptual framework, a tool to help facilitate understanding of the relationships among key variables in relation to the real world. To illustrate the pressures surrounding where new trees might go, I thought an updated version could look something like this:

> Land for conservation of primary ecosystems, including forests = Total land area – (Agricultural area + Restored or renovated forests + Settlements)

Any increase in restored or renovated forests adds more to the area on the right side of the equation unless it's occurring on already-protected lands. The constraints highlight the pressing question: Where do the new trees go?[8]

Jad said that at the COPs, everyone talks too much about *forests* and not about *forestry*. "Forestry is where the solutions are. What are we going to plant? Are we using climate-informed reforestation techniques to ensure these forests are going to survive? Those are forestry questions that would benefit from global collaboration."

They are also questions about economics and current land use and food production. There are very practical considerations about *how to do* the forestry and *who leads* each effort, even if the potential is there and the financial support is increasing for a forested future.

The Land Gap Report claimed that keeping existing ecosystems healthy and functional is the most important land contribution to limiting global warming. Protecting and restoring existing forests should be the top priority to maintain the stable carbon that they already store and to facilitate more carbon sequestration over time. It argued that planting trees to create more forest cover "should not compete" with other important land uses. There's a difference between planting trees on lands that have been degraded or are not currently productive and planting on lands that people are using to produce food or provide other services.

Listening to Jad and imagining that formal room filled with government leaders and eager observers from organizations around the world brought me back to Rachael Garrett in Zurich, to that little corner café on a loud city street not far from Tom Crowther's office. We had been talking about a more holistic vision of justice-centered restoration—an approach that returns ecological functions to a landscape but also supports the people who live there and includes them, democratically, in the endeavor.

Targets and commitments are top-down. The actual doing has to come from people in place. *Is this movement for the people? Or by the people?* I wondered. If it's both, which I think it should be, how can it include everyone who needs to be involved over time—tree time?

Rachael had talked about Wangari Maathai's *Replenishing the Earth*, her 2010 account of the Green Belt Movement and the underlying values that enabled action. It opens with a description of how people can better understand their environment as part of their community and become empowered through restoration. In Maathai's eyes, the movement's values were love for the environment, self-betterment, gratitude and respect, and a commitment to service.

"She talks about how when she started this restorative process in Kenya, she felt like so many people engaged with it well because they finally had some power to take control over what was happening around them and to make it better," Rachael told me at the café. She said the "beautiful seed" of this approach to restoration was there so long ago, and "we've kind of lost it."

Maathai was writing over a decade before the Global Potential study, the flurry of media, and the misunderstanding that planting trees could be the silver bullet for halting the climate crisis. The contrast in motivations made me think about how focusing on numbers of trees and the statistics on potential carbon alone can overlook what real restorative solutions look like on the ground and in communities. I also thought of the version I'd read with Calder and my doubts that every planting effort underway could be as holistic as Maathai's movement in Kenya or that the exact same strategy could be implemented elsewhere.

"How can you reconcile a global map with pixels that are appropriate for restoration with this grassroots vision of people actually being able to take control of their own environment and make it better and restore its functioning?" Rachael said. "I am not saying we need to do away with these global maps, but every time scientists make those, they also need to draw attention to these other processes that don't make it into *Science*."

She was referring to the social, economic, and even ecological processes that support any sort of reversal from forest loss to gain that endures across tree time. As ecologists have written and more people are now discovering, the gain is not a mirror image of what was lost. In most cases, destruction of forests is generally concentrated and abrupt; Virginia Norwood's multispectral scanner system sensor and all the satellites and tools that came afterward have been very good at tracking loss and degradation with imagery in sequences. But any gain is more dispersed and variable; it's impossible to track in the same way. Any gain requires far more commitment over time to sustain the change.[9]

I could see those stepping-stones at Kew Gardens: What—do—plants—need—to—grow?

"What do they need to grow, Mama?"

Perhaps I never answered Calder properly that day because I felt ashamed of how complicated growing trees, in particular, has become, even in a world with more and more people promising a better future for forests.

On a subzero January day in Bozeman, I was talking with Bethanie Walder at the Society for Ecological Restoration (SER). SER is focused on not just forests but all ecosystems; it is heavily involved in the UN Decade on Ecosystem Restoration and has helped develop the central principles guiding the global effort. Bethanie is passionate about restoration, and she is generally very concerned about the misconception that just planting trees can restore ecosystems. On that afternoon, we were discussing the differences between reforestation and restoration.[10]

"When we talk about reforestation and the science behind reforestation and how it plugs into the carbon arena, I think we're approaching it from the wrong place," she told me emphatically. "Carbon is a product. It's not the sole purpose. And if restoration is the purpose, and that's what the UN Decade on Ecosystem Restoration is about, then we can have a *lot* of benefits. Carbon is one only of them."

She referenced what SER calls the *restorative continuum*—a framework for thinking about a range of activities that can improve environmental conditions and reverse degradation. On one end of the spectrum is simply reducing negative impacts to society; on the other end is fully recovering native ecosystems. Steps in between include, for example, repairing certain functions of ecosystems or initiating native recovery.

"The farther to the right you are on the restorative continuum, the greater the biodiversity, human health and well-being and climate benefits from the restoration activity," Bethanie said. "If the only objective is tree planting for carbon, then the project may not even qualify as restorative; it wouldn't even be on the continuum if it causes collateral damage like biodiversity degradation." At the upper end to the right is what Bethanie also referred to as "five-star restoration." She and the ecologists at SER navigate any renovation/restoration debate with an understanding that ecological restoration aims to remove degradation and assist in recovering the trajectory an ecosystem *would* have been on if the degradation had not occurred. It's going forward to what's possible. They add the caveat that the trajectory must account for environmental changes, including those driven by climate. Even five-star restoration is not a mirror image of what once was.

Bethanie had taken an unlikely route to restoration. She'd spent nearly two decades working to reduce the impacts of roads (especially

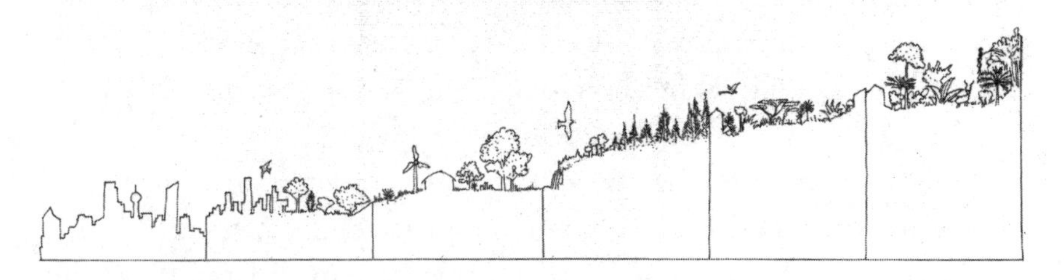

The restorative continuum with reducing societal impacts on the far left and fully recovering native ecosystems on the far right.

Credit: Lorenzo José Rosenzweig, adapted from figure presented in Gann, George D., Tein McDonald, Bethanie Walder, James Aronson, Cara R. Nelson, Justin Jonson, James G. Hallett, et al. "International Principles and Standards for the Practice of Ecological Restoration." Restoration Ecology *27, no. S1 (2019): S1–46.*

dirt roads) on many kinds of ecosystems, including forests. For dirt roads that are no longer needed, the best way to address their detrimental impacts to the environment is to physically remove them. "You could walk an abandoned road that has twenty-year-old trees growing on it," she said. "But if you dug into that soil, you'd see that you have water only infiltrating the first twenty centimeters, or so. It was *revegetated*, but not restored. There's minimal soil carbon. It's nitrogen depleted. It's essentially dead soil underneath. For all intents and purposes, it *looks* like it's been restored, but it is not." There was a researcher, Rebecca Lloyd, who studied how forest ecosystems recover after the abandonment or removal of roads in Idaho. Removing the roads and recontouring the land's surface, instead of just abandoning the roads and allowing plants to return at their own pace, put the soils on track toward restoration by yielding more carbon, more nitrogen, more organic matter over time.[11]

"You could drive a bulldozer up to a forty-year-old tree that's growing in functional, living soil, tap the tree with the bulldozer, and your bulldozer will go backwards. You do that on an abandoned road and the tree will fall over, because its root layer is some twenty centimeters deep. That's it."

She was fired up, speaking quickly. "You could measure that site with satellites, and sure, you've got reforestation. It's beautiful, but you may also have totally dead soil. You don't have carbon sequestration happening in your dead soils. It's not *restored*. It's *revegetated*. That's different." Bethanie paused and exhaled, slightly exasperated by any misperception of the two as being equal. "There's this overreliance on satellite vegetation data that misses a lot of information. Everyone wants simple answers and ways to track progress, but restoration is really complex."

SER created an ecological recovery wheel, a visual aid and tool to assess progress toward full restoration over time. It shows the diversity of targets for recovery—from species composition to structural diversity, absence of threats, and more. The idea is that someone (or a team) tracking a project and using various indicators or informal observations can also score its condition and recovery progress. As I spent time considering all that the figure encompasses, what became most evident was how the process and goals of restoration contrast with those of planting trees for climate mitigation. If you're focusing on one outcome—the amount of carbon sequestered by trees over time—then counting trees, monitoring tree survival, and measuring biomass to estimate carbon seems like a logical approach. The problem is that maximizing for any one thing—be it carbon or any other "product"—at a large scale, as fast as possible, typically has some other consequence.[12]

Most people I'd interviewed had said forest restoration can fall under the umbrella of reforestation; *forest restoration* is bringing back trees to a landscape in an ecologically restorative and sustainable way that takes time and steady care. Bethanie argued that restoration often includes some reforestation, but a lot of other kinds of reforestation should just be called something else completely.

"When I hear the term *reforestation*," she told me, "I assume it means the re-creation of a forest. Therefore, a monoculture tree plantation doesn't fall under reforestation. But other people hear reforestation and assume it means any manner of replanting trees, so a monoculture tree plantation can be reforestation."

She was working off a definition of forest that is far from just tree cover. There are still forests that deliver some modern semblance of what they once did before the industrial age. Then there are the forests designed to deliver on a single value—be it timber or carbon, for example; these forests are farming for specific products. I knew her reluctance to call a timber plantation a "forest" would outrage some foresters, but it also brought

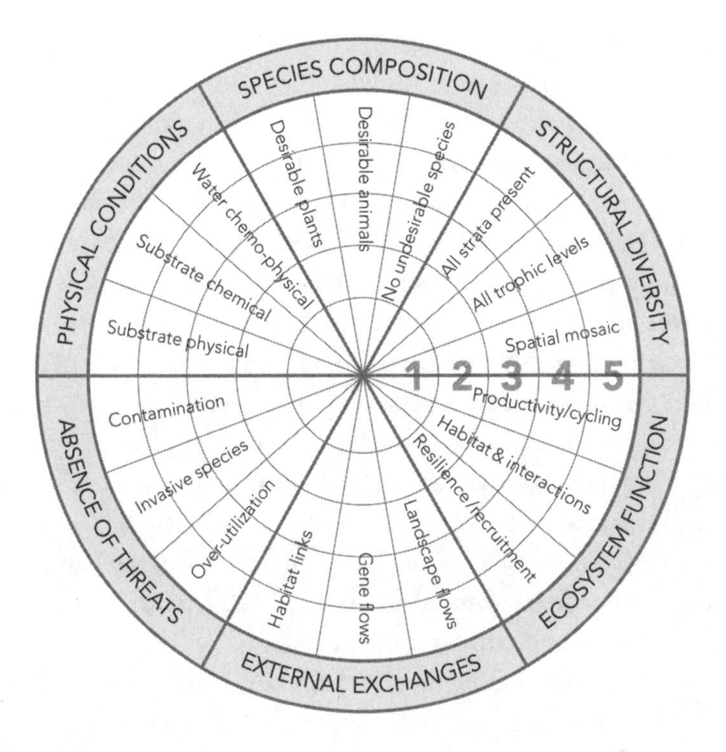

The ecological recovery wheel.

Credit: Gann, George D., Tein McDonald, Bethanie Walder, James Aronson, Cara R. Nelson, Justin Jonson, James G. Hallett, et al. "International Principles and Standards for the Practice of Ecological Restoration." Restoration Ecology *27, no. S1 (2019): S1–46.*

home the point that any interpretation of what a forest is influences what we may try to protect, foster, or re-create.

Eric Lambin and Patrick Meyfroidt shared a story with me from their work on forest transitions that reveals yet another layer of complexity. They began working together not long after Alexander Mather's research on forest transitions in Scotland and Europe had caught Eric's attention and he'd met Mather in Aberdeen. This was back in the early 2000s, as Eric was shifting from a focus on the causes of deforestation and becoming interested in unpacking the drivers of forest gain at national levels in countries experiencing forest transitions. Patrick, like Eric, was drawn to the places where positive change was occurring, where people's destruction gave way to efforts to re-create what they'd lost.

"We started with Vietnam," Eric explained in a talk at Stanford. "If you look at the curve in Vietnam, you see forest cover going down for fifty years. And then there's this turning point, really sharp from one year to the other. They just go to massive reforestation with a rate of reforestation twice as high as the rate of deforestation before." That turning point for forest cover in Vietnam had occurred in the first half of the 1990s.[13]

Patrick began pursuing a PhD in 2004, and by September of that year, he had his eye on Vietnam's reversal and what had caused it. Fast-forward a few years, and he and Eric found that the development of markets for agricultural goods had increased agricultural productivity in mountain paddies and maize fields; the shift allowed for reforestation to occur on the more marginal hillsides. They'd documented a logging ban as well as other policies to address forest degradation. Their first study of the causes of reforestation in Vietnam came out in 2008.[14]

It was also clear that plantations were a big part of the story.[15] The increased tree planting was associated with a decline in natural forest cover. Eric told me,

> About fifty percent was plantation and the other fifty was natural regeneration of forests. So, clearly, we understood that the definition of forest was unsatisfactory. If you take FAO's [Food and Agriculture Organization] definition—you know, anything with more than 10 percent of tree cover that is any vegetation taller than five meters over a certain area—that will include oil palm trees and eucalyptus and the like. And very, very quickly, you know, the ecologists were kind of unhappy because they said, "Well, you are mixing up really biodiverse forests with monocultures of exotic species."

A Dutch scientist, Meine van Noordwijk, started referring to *tree*-cover transitions instead of *forest*-cover transitions around that time because, as Eric described, "What we were seeing was more trees on the landscape; not real forests." (van Noordwijk also called the term *forest* an "institutional marker"; as a policy brief stated, "There are trees without forests and trees outside forests.") If there's a more widely accepted descriptor for any and all kinds of forest gain—from afforestation (establishing a forest on land that was not recently forest) to reforestation, agroforestry, or forest restoration—it's "increasing tree cover."[16]

Then one day Eric, reading the newspaper in a café in Louvain, came across a headline about exploding exports of wooden furniture in Vietnam.[17]

"What?" he thought. "How is this possible? How could they export so much wooden furniture when they'd protected their forests?" Patrick led the effort to investigate further, which revealed a massive amount of trade from Laos and Cambodia, the neighboring countries, into Vietnam. Vietnam was importing and processing the wood and then exporting it as furniture to the United States, China, and Europe.

"So, in effect, they were just offshoring their land use, their deforestation, to their neighbors."

The phenomenon is called *land-use leakage*: an environmental policy in one place leads to a change in land use in another that then reduces the overall effectiveness of the initial intervention. It's like offsetting a positive action in one place with a negative action in another, and it's difficult to track. Solving a local problem can displace the problem elsewhere in the world.[18]

"We thought maybe we had this weird outlier," Eric shared with me, when he and I talked further about that moment of realization and all the research that came afterward. "But it was not an outlier." Patrick, Eric, and Tom Rudel (the scientist who had characterized this current movement as "something new under the sun") went out to investigate leakage in developing countries that had recently experienced a forest transition—Bhutan, Costa Rica, El Salvador, Chile, India, and Vietnam—and one developed country, France, which experienced a transition during the nineteenth century.[19]

"They were all doing it, and actually Vietnam was offshoring less than the others."

When I was talking with Eric, I called the leakage issue a *problem*; he paused on that and explained that it took them another five years of research into other cases to realize it may not always be a problem, per se. Bhutan, for example, has imported wood and charcoal from plantations that are generally well managed in India.

"It's much better for Bhutan to get its wood and charcoal from these plantations in India than from its biodiverse primary forests." In 2011, 68 percent of the total forest area required to produce wood consumed in Bhutan was in India.[20]

Oof, talk about complicated! It's mind-bending to try sorting out "the good" from "the bad" and what delineates the two.

As Eric noted, if countries with tropical forests that have a lot of biodiversity and store a lot of carbon then import wood from Scandinavia or Canada, where existing secondary forests can supply timber, that's probably a good thing. Ideally, it avoids degrading more intact areas. The land use for timber shifts elsewhere, thus negating the intervention from a global perspective, but the leakage across borders could have other local benefits for the different services that forests provide.

Costa Rica is another example of this. Between 1985 and 2013, the United States was the largest supplier of wood products to Costa Rica. Cheaper pine imports came later from Chile; much of that wood went toward pallets that were then used for exporting bananas and other crops. There were policies restricting forest clearing for agricultural expansion, but Costa Rica didn't have a policy that directly swapped forest protection and restoration in one place for harvests in another. It just happened: the demand was there, so the wood would come from somewhere. Some scientists call this *benign leakage*. Patrick prefers *spillover* as a neutral term, and even *positive spillover* when there are beneficial outcomes on the same variable. Halting the clearing of *primary* forest in one place hadn't caused clearing of primary forest in another. Instead, Costa Rica had "displaced land use" of its relatively intact tropical forests to *secondary* forests used for timber in the temperate region of the United States.[21]

"That was probably a good thing, again, because the biodiversity in Costa Rica is quite unique," Eric said.

Moreover, the carbon sequestered in the Costa Rican forests likely outweighed the carbon consequences of harvesting the US forests. Tracking displacement of the loss of forest (and of what kind of forest) is one hurdle to comparing the net outcomes of such spillovers. Then trying to determine how the displacement affects other values, such as biodiversity or carbon sequestered, is even harder. There are other factors to consider in the carbon accounting, such as the emissions associated with transporting the timber from one country to another.

I went to Patrick to ask how the lessons from his, Eric's, and others' work on leakage might relate to the reforestation efforts today. Would growing more trees in some places just increase the pressure to deforest elsewhere?

"If you have small-scale reforestation, it's unlikely to create a lot of leakage," Patrick told me. "You reforest by starting with the low-hanging fruits and places with low-opportunity-cost, marginal lands." There is relatively less pressure in those places to put that land to a different use. If you're then able to ramp up reforestation, coupled with forest protection, to a global scale, you're unlikely to have a lot of leakage either. That's because in an ideal world of well-functioning markets and political institutions, land-use allocation would be coordinated at the global scale. The argument hinges on a widespread revolution in land-use and food systems, such that keeping forests is economically viable and making agricultural practices more intensive and sustainable helps spare land from more clearing. It hinges on actions big and small: farmers adding trees to their properties, for example, and big corporations supporting better practices along their supply chains to affect a lot of land. This would be a new functioning system. "It's in between where there is this bottleneck," Patrick explained. "When you start to scale up reforestation, tree plantations become a significant use of land, which has large impacts on other kinds of land use like food production, nature conservation, and so on. There is a bottleneck stage where leakage matters, due to competition in land use and national and subnational jurisdictions." That's the danger zone, where complex spillover and leakage dynamics arise because some forests are protected but others are not, and some reforestation programs might be implemented without concern for food security elsewhere, for example.[22]

He added, "You might expect that at the end of the day, when you really have all countries on board and everything operating well under this global system, then leakage wouldn't be a problem anymore." If all countries are implementing effective forest conservation and reforestation interventions, there wouldn't be room for deforestation.

I thought about the Declaration on Forests and Land Use—signed by 141 countries representing 90 percent of the forest cover in the world. *Is that enough? Is everyone really on board?*

"But that would be in theory," Patrick added. "I would say that in practice, solving one problem often creates others."

In other words, a haphazard medium-scale movement to reforest parts of the planet can result in unintended consequences. If growing more forests triggers leakage of agricultural practices into other natural ecosystems, unexpected impacts will occur. Furthermore, even an effective movement to grow more forests and curb agricultural expansion could lead to new pressures on the environment. He mentioned the increased demand for rare earth metals for use in renewable energies, such as solar panels, as a general example of one good causing another bad. While the renewable energy reduces fossil fuel emissions, the mining creates other problems. Only so much swapping can occur between high-integrity forests in some places and "tree cover" in others before the balance of what those natural, functioning, or even ecologically restored forests provide and what tree cover might deliver gets thrown off-kilter. As Patrick also noted, even at a global scale, it's possible that such a big wave of action to reforest could result in a new set of problems that no one can entirely predict.

I tracked down a copy of Maathai's book soon after meeting Rachael in Zurich and quickly lost myself, late at night, with a cup of tea in her moving account of reviving the land and bringing back trees. The benefits of such a community-led effort contrasted with the concerns that Patrick had raised about attempting similar work at a great global scale.

"After a few years," she writes, "I came to recognize that our efforts weren't only about planting trees, but were also about sowing seeds of a different sort—the ones necessary to heal the wounds inflicted on communities. . . . What became clear was that individuals within these communities had to rediscover their authentic voice and speak out on behalf of their rights (human, environmental, civic, and political). Our task also became to expand democratic space in which ordinary citizens could make decisions on their own behalf to benefit themselves, their community, their country, and the environment that sustains them."[23]

Maathai describes our damaged world as impacting not only the environment in which we live but also what "we might call our inner ecology, our soul, and sense of being human." I thought of Calder and his instinct to make friends with trees, to name them and care for them, as well as of the studies on how "exposure" to forests can affect human well-being in a variety of ways. Trying to bring back a forest or restore what's degraded doesn't need to be about targeting just one outcome—such as carbon sequestration. "In the process of helping the earth to heal," Maathai notes, "we help ourselves"—probably in more ways than anyone can measure.

Persistent patterns of destruction, Maathai explains, are based on behaviors that "may have been suitable for us when we were fewer in number and could destroy vegetation and move on with relatively limited effect on the environment."

I read this part over a few times, taking in what the warming world demands: "Climate change is forcing all of us—rich and poor—to acknowledge that we have reached a point in the evolution of this planet where our needs and wants are outstripping the ability of the earth to provide, and that some of us will have to do with less if those who have very little are going to have enough to survive. It may require a conscious act of some of us saying no in addition to finding other, less destructive ways to say yes."

Fix something locally, like forest loss, and it leaks to some other geography. Try to fix forest loss globally, and eventually the consumption issue must be addressed. There's only so much land. How do all these countries making commitments find the path together toward a less destructive way of saying yes?

"If we're lucky, the seeds we plant will germinate and grow, which is an added bonus we can enjoy and be encouraged by. But genuine change is rarely easy," Maathai writes. "Quite often, our work is a struggle."

It turns out that Virginia Norwood's granddaughter Rivah and her husband, Johann, own a farm in Hawai'i, about two miles from where I stayed in the yurt. I went to meet them after my visit with Terraformation and before I left the Big Island. There were boxes of things left unpacked in various rooms. I hadn't noted this as strange until Rivah pointed them out.

"We've been here a couple years, but the inside of our home doesn't reflect that. We have been so focused on outside, on restoring the soils and making the land green again. For us, that has come first."

I had never considered what my own fairy tale version of restoration could be until I set foot on their property. I quickly realized that Rivah and Johann were living it with a beautiful farm on a hilltop with ocean views. They were raising a kid with fresh fruit to pick outside the front door. They were working hard to care for the trees and the soil on their property. They'd moved there from Miami, where they'd finished doctoral programs, Rivah in coral biology and Johann in marine ecosystems and society. Some years back, they'd visited a farm in Georgia that was promoting a negative carbon footprint, achieved through grazing rotations on pastureland that facilitated more carbon sequestration than conventional beef production.* A documentary film, *The Biggest Little Farm*, had also inspired them to overcome the inertia of their daily lives. It was about a couple that left Los Angeles for a 200-acre farm in the foothills of Ventura County. The land had been nutrient depleted and drought stricken, and the couple had brought it back to life.[24]

"We'd bought their products at the farmers market, not far from where Naomi and Virginia live," Johann said.

Rivah described their own leap to purchase land in Hawai'i as going into "eco-prepper mode." They wanted to find a place where they could live self-contained and sustainably off their own land.

"We wanted to live in a way that felt more aligned with our values," Johann said. "I wanted something that I could touch again, instead of sitting in an office in Miami with the air conditioning cranking, trying to do conservation work for a place 5,000 miles away." We were standing around their kitchen island while their son, Callan, babbled and played at our feet. They were both still working remotely for other

* A group of scientists later conducted a full life cycle analysis at the farm, White Oak Pastures. The findings confirmed that multispecies pasture rotations sequestered enough carbon in the soil to create a greenhouse gas footprint that was 66 percent lower than that for conventional, commodity production of beef. The problem, however, was that the regenerative approach required 2.5 times more land. (Rowntree, Jason E., Paige L. Stanley, Isabella C. F. Maciel, Mariko Thorbecke, Steven T. Rosenzweig, Dennis W. Hancock, Aidee Guzman, et al. "Ecosystem Impacts and Productive Capacity of a Multi-species Pastured Livestock System." *Frontiers in Sustainable Food Systems* 4 [2020]. https://www.frontiersin.org/articles/10.3389/fsufs.2020.544984.)

efforts—Rivah designing exhibits for museums around the country and Johann running a conservation organization in the Galapagos. But I could see and feel the attraction of the work they'd chosen in place and at home. In a couple of years, they'd removed invasive species from the overgrown land, planted more trees, replenished the soil with nutrients from compost and trimmings from the property, and started harvesting the fruits of their labor.

They'd planted about 500 trees already, and they had plans to add another 500 in the next year. "They're fruit trees, so that we can produce more. Sure, they have a carbon side effect, but carbon wasn't the purpose."

"The big old *ōhi'a* are still there," Rivah added, "and some of the mid-story as well. He was referring to the layer of trees and vegetation between the low-growing plants and the overstory trees. But as we go in and remove the invasives like the strawberry guava, we're discovering there's more native species in the forest, waiting for light. They pop up when given the chance." Rivah and Johann were more focused on the soils and the health of the whole system; trees, whether fruit trees or other native trees, are just a part of that.

"Most industrial agriculture soils have been reduced to about 1 percent carbon, which is why we have the Dust Bowl in the Midwest," Johann said. "They are basically left with dirt, not soil, only silica particles and no humus in it anymore. It can't hold water, which is why wind carries it away and creates dust storms. That happens when you only have 1 percent of the carbon left in the soil, whereas a forest floor or a functioning pasture would have more like 5 percent." He'd read that getting soils globally to just 3 percent could sequester a terratonne of carbon, which is nearly five times the 200-gigatonne potential for trees from the Global Potential study.[25]

"So, the solution—in my head—is natural. I look around now, and there's so much biomass we've created. It's astonishing how quickly this place has come back to life."

"When we first came here," Rivah added, "there were some birds, but there wasn't this cacophonous dawn chorus. Now the insects and the birds are back. It's been this dramatic population explosion; we find nests all the time."

Another study found that natural areas store more carbon than farms. The authors suggested that there could be a counterproductive transformation of natural habitat to agricultural land when trying to increase carbon storage. This is because they found lower agricultural yields to be associated with higher above-ground carbon stocks at some sites, such as in Ghana and Mexico. Lowering productivity of farms, however, could create pressure for agricultural expansion elsewhere—another example of leakage.[26]

Rivah loaded Callan into a wrap to carry him on her chest, and we set out together to see the property. The boundary between farm and forest fell where the land became steeper; they'd been working just as hard to care for the native trees as the fruit-bearing ones below. There were mounds of woodchips from invasives they'd cleared and compost piles to enrich the soils. I noticed the yellowed leaves of nutrient-depleted trees on the neighboring property, but the leaves on their trees were dark green and flourishing. Johann grabbed an orange from a limb above us; I could smell it before I held it.

Their version of restoration was the most romantic one I'd encountered yet because it was about working with each other and with nature in their own backyard. Sure, there were technological aspects to what they were doing: farming tools, harvesting equipment, a buggy to get around the property, and a water catchment system with storage tanks and pumps. They had aspirations for solar power. But above all, I could see they were doing their best to revitalize and become a part of the ecology of a place instead of just taking from it.

"Making agriculture regenerative is a solution that is readily available, scalable, and technologically possible, with a workforce already in place to get the job done," Johann explained. His commitment and enthusiasm reminded me of Yee Lee's perspective on restoration and growing more trees. "The solution is right in front of us," echoed in my head.

As we stood outside together in the sun below the *ōhi'a* trees, I tried to be fully present as I learned about the relationship they were creating with their land. But I also couldn't stop thinking about scale.

"We're not billionaires," Johann had said in the kitchen. "Maybe a billionaire can have a big impact at a big scale. But we are many; they are few. If we all contributed small solutions, it would add up to something much greater than a few big solutions."

Was he just an idealist? I wondered. Or was this Wangari Maathai's lesson: all the high-level commitments for tree planting and restoration in the world will not matter unless local people can act upon them and then sustain them. Even then, avoiding the bottleneck when so many projects are happening so quickly seems impossible; there's bound to be some messy stage when the effort to resolve one problem creates others.

Successfully implementing the Bonn Challenge, which aims to bring 350 million hectares of degraded or deforested land into restoration by 2030, would be the most rapid and intense land-use change in human history. There are more than 200 million hectares of maize, or corn, one of the crops that uses the most land in the world. However, it took 10,000 years to reach that scale after humans first domesticated maize. The rapid expansion of soy farms transformed the Amazon, but it took decades for that to occur. The UN Decade on Ecosystem Restoration echoes that restorative mission and moves beyond forests to include other degraded ecosystems too. The Forests and Climate Leadership Partnership became another ambitious step forward for forests. They all set high-level goals to combat the social problems connected to the ecological ones at a fast pace and to address the climate crisis too.[27]

I kept reflecting on Maathai's motivating values and the effort in Kenya, about Rivah and Johann's home-based approach as well. Perhaps one simple big idea for the world might be the biggest fairy tale of all. I thought there must be a way to bridge the gap between restoration in its most romantic and holistic form and the large scale needed for a global impact.

How?

11

ALL HANDS ON DECK

Why would companies be interested and involved in this?" Dominic Waughray, a managing member of the board of the World Economic Forum posed the question to Marc Benioff, the chairman, cofounder, and CEO of the software company Salesforce. "It sounds like this should be something for NGOs and governments, not companies," he added.

Iván Duque, the former president of Colombia, was sitting between them; they were all dressed sharply in business suits for the press conference. Sadhguru, founder of the Isha Foundation, an ashram and yoga center in India that leads educational and spiritual activities, sat on the other side of Benioff. Jane Goodall was next, her straight, white hair pulled back into a ponytail. A necklace with a pendant of Africa dangled over her turtleneck sweater. Hindou Oumarou Ibrahim, president of the Association for Indigenous Women and Peoples of Chad, was at the far end of the table; in 2019 *Time* magazine had listed her as one of fifteen women championing action on climate change. An orange, flowered *lafai* was wrapped around her head. They were at the World Economic Forum in Davos, Switzerland—January 24, 2020.

"Well, we are certainly at a time of planetary emergency," Benioff replied. "We realize our planet is getting warmer, and we need to find ways and solutions that we are going to all become, number 1: carbon net zero." He was referring to the state where the amount of carbon released into the atmosphere is balanced by the amount removed. "And number 2, we have to sequester or eliminate the carbon that has been emitted into the atmosphere. Human beings are very good at moving carbon around, and we can see what we've done. We've moved it from the ground and into the atmosphere, from the ocean into the atmosphere, from the trees and into the atmosphere. We need to get a big eye on our forests, so we can use them to sequester this carbon."

Benioff described a meeting in Geneva in August 2019 when former vice president of the United States Al Gore, who has been an advocate of climate action for decades, mentioned the Global Potential study. Benioff hadn't seen it yet and replied, "What? One trillion trees will sequester more than 200 gigatonnes of carbon? We have to get on this right now! Who's working on this?"

He thanked President Duque for his leadership in what became the "1t.org" campaign for a trillion trees and, almost in the same breath, President Donald Trump for his commitment to participating.

"I know that planting a trillion trees sounds daunting, but there used to be six trillion trees on Earth. Now there is three trillion. So, we should be able to get another one trillion back. Well, I'd like two trillion. I guess I want them all back, Tom!" Benioff declared, gesturing to Tom Crowther, who was sitting in the front row of the audience, inches from Goodall and the eclectic group of panelists.[1]

"How about all six trillion?" he added. From my perspective, this was another watershed moment that helped broaden support from the private sector for the movement already underway.

When I'd met with Tom in Zurich, he'd told me about the time he first met Goodall in Davos. "So, you're the scientist who led the study about trees?" she'd said. "Well, some people unfortunately thought it was just about planting trees," he'd responded, then added something along the lines of "and it pissed a lot of people off." He had been referring to the drama surrounding misperceptions of planting trees as *the* solution for climate change and everything that overlooked. She'd smiled and clarified that he couldn't have gotten there without ruffling some feathers. I loved the response from such a conservation heroine; any big advance or discovery comes with people misunderstanding or getting upset for one reason or another.

Benioff said they'd already seen interest from 200 to 300 companies within twenty-four hours of announcing the campaign that was created to serve the global community of actors committed to growing, restoring, and conserving a trillion trees worldwide by 2030. (Note: 1t.org wasn't focusing only on planting, but broader perceptions of and emphasis on planting as the primary activity still prevailed for some time. Behind the scenes and before Davos, there were other organizations pushing for recognition that the one trillion figure would include *conserving* trees—that is, avoiding their loss and continuing to support their growth—not just planting new ones.)

"Who's against the trees? I mean, everyone is for the trees!" he said. "The trees are a bipartisan issue. Everybody is pro-tree. I haven't met any anti-tree people yet." In a subtle bow to her years of environmental advocacy, he said Goodall had done a great job.

"Every one of us can join this movement, and every one of us can make a commitment to plant trees," Benioff added.

When it was Hindou Oumarou Ibrahim's turn, she began, "For Indigenous peoples, forest trees are not only a ton of tree or leaf. It's our home. It's our food. It's our medicine. It's our life! It is our knowledge. It is our school. So, when we hear about the initiative of restoring the forest ecosystem, planting a trillion of trees, we are the first ones who are very excited about it."

Sadhguru said it was exciting that the world was starting to think in terms of trees, but he also seemed less starry-eyed than realistic about the endeavor and timeline. He called the trillion trees "an inspiring number to go for" but noted that we don't typically plant trees. "We either plant seeds or saplings. Before we call it a tree, it will take six to ten years. That's when it becomes a tree." Sadhguru stressed the importance of bringing trees back to farms.

"This is what we have done in southern India, where right now there is a project to involve family and farmers to grow timber in their lands as part of agroforestry, which includes other crops in between. It is tree cultivation with crops, and crops come out much better. The nutritional values are better. The soil recovers. The water tables come up, and overall there is a long-term crop on the land." Sadhguru noted that in the next decade, more than 200 million people in India will migrate to cities that aren't prepared for the influx. But when farmers have long-term crops on the land, they won't leave their land for the cities. Then, there is a standing investment: "Money is growing on the land." He called this an economic process for the people.

Another point he made brought me back to the notion of ecological baselines. *What is my baseline? What is my son's? What is yours?* In places where trees are long gone, he suggested, people lose their connection to them. *Tree* becomes only a word in people's mind.

He had trained a few thousand people to renew this connection to trees and to teach others to do the same. "We made people sit with eyes closed, and made them breathe, and set up a certain process where your exhalation is tree's inhalation, tree's exhalation is your inhalation. Once they felt this, you [couldn't] stop them from planting trees, because one half of your breathing apparatus is hanging up there on the tree." Maintaining this connection is also necessary, he said, because it takes ten or fifteen years for a tree to get to a reasonable size, but it can be cut and taken away in one afternoon.

Sure, there's the global interest in carbon sequestration, but it's the local benefits and support that can sustain care across tree time.

"We never talk to the farmer about saving the planet. It is just that to enhance his economic condition, this [planting] is being done. If we do not marry ecology and economy, economy will win hands down." *The connection to trees can only go so far to influence human behavior,* I thought. There must be a pathway that is also economically viable for change in land use to occur.

Kevin O'Hara became the lead of the US chapter of 1t.org in August 2021; he's helped galvanize forward-looking companies that care about forests and climate in the United States to make pledges. At the Davos press conference, Benioff had said he'd expected about fifty billion to one hundred billion trees to come from the United States. By the time I interviewed Kevin three years later, the pledges had crossed into that range. Although those pledges came from companies, state and local governments, and other organizations within the United States, the trees could be planted in locations across the world. Another seventy billion trees came from pledges made in China.

"At the US chapter, we've got 100 pledges in the ballpark of fifty-five billion trees," he said, "but we have no clue of what that means on the ground, because it's a self-reporting mechanism." A company, corporation, or nonprofit makes a pledge of a certain number of trees, hectares, or acres restored or protected and of other supporting actions that make the work possible, and they're also the ones to submit annual progress reports. He called the platform a "catcher's mitt for ambition."

"In the dugout then, everyone is sitting and thinking about what are the plays we can make? What are the partnerships we can create and the funding sources that we can catalyze? How do we pilot and share best practices?"

I asked Kevin what he thought about the high-level commitments, given that the actual work and on-the-ground outcomes occur in local communities and on local lands.

"It's a top-down movement with these targets, but then the solutions have to come bottom-up, right?" I inquired.

I don't like using that language of *top* for leaders and *bottom* for people like me or any other citizen of any country, but I didn't know what other words to use. Kevin said the movement was both, coming from leaders and from communities themselves. Framing this dichotomy as structural versus individual forces of change is another helpful way to think about the drivers; the structural forces come from new policies or economic approaches to valuing nature, for example, and the individual ones come from the actions people take.

Kevin offered examples to show the diversity of projects out there: US pledges have included the City of Detroit planting street trees, for example, and efforts by major organizations like Eden Reforestation that have supported large-scale reforestation in the Amazon or elsewhere.

"I think it's helpful for local groups to have something bigger to be a part of. Say, you're Casey Trees in Washington, DC, and you're doing all this great work to try to reforest a city. You've got your citywide goal of increasing tree cover from 30 percent to 40 percent, and you're working so hard on those goals." Casey Trees happened to be another organization I'd heard about when I was in Seattle for the urban forestry conference. "But what does that lead up to?" he asked. "We're doing this for all kinds of great reasons, yet part of the ambition piece is to feel that you are contributing to something bigger than just your community, just your company, or just your state. It's knowing this is an all-hands-on-deck moment for our country, for our world, as far as climate. We know that addressing climate change is a big piece of this work, but it's also for health, for water, for biodiversity, for recreation, and more."[2]

The phrase "all-hands-on-deck moment" echoed in my head. *So true.* It spoke to the race and to the hope I felt and still feel.

Russia, the United States, Canada, Australia, Brazil, and China are among countries that have the most land available; flag their potential. If you're talking *just* carbon, Central Africa is another region to prioritize; land has some of the highest potential for carbon accumulation per hectare; scientists also predict forests in that region to be quite resilient to future climate conditions. But "all hands on deck" speaks to the fact that there isn't and will never be one way to solve the climate crisis or one way to protect forests or restore tree cover. In that moment with Kevin, I felt oddly self-conscious about the limited number of projects that I could include in my writing; any in-depth portrait I offer of an effort to shift forest loss to forest gain would be a tiny window.

I took "all hands on deck" to mean that we all must try, and there are many ways of trying. The alternative of giving up and doing nothing isn't an option; it threatens our survival.

Before there was 1t.org, there was Trillion Trees. (I know: keeping straight all the targets, commitments, platforms, and organizations can get overwhelming.) The timing of the two efforts is relevant, because after its founding in 2016, Trillion Trees, a consortium of the Wildlife Conservation Society, BirdLife International, and the World Wide Fund for Nature UK, spent the first years focused more on protecting forests and halting deforestation than on restoring forests. Tom Crowther's earlier tree count—three trillion trees in 2015; six trillion at the dawn of civilization—had been a spark for the effort, amidst heightening global concerns about forest clearing and climate change. Yet the initial vision document for Trillion Trees, released in 2017, separated a trillion trees into three categories: trees regrown, trees saved from loss (meaning avoided deforestation), and trees in areas requiring increased protection, a category at least as big as the other two combined. That breakdown had been supported by another study that Tom had led; it estimated that only about 590 million trees could come from *new* trees, which represented a maximized scenario. Clearly, Tom and the members of the Trillion Trees consortium weren't *only* thinking about tree planting.[3]

As one staff member at Trillion Trees shared with me, "Of course, 'our trillion trees' never planned to *plant* a trillion trees; it's always been about ensuring one trillion more trees on the planet than would have been with business as usual. So, that includes trees protected, saved from loss, and restored."

Yet, when 1t.org launched in the wake of the Global Potential study and all the media surrounding tree planting, a call to action for planting trees was more tractable than the typical cries for halting deforestation or increasing protections for existing forests. By 2023, Trillion Trees wasn't splicing and dicing a trillion trees into categories anymore or pushing 1t.org to do so either. (I get it—why complicate things further?) As for curtailing emissions or deforestation or even achieving other global targets, when have we ever hit any target right on schedule? Staff members at Trillion Trees told me 1t.org became a rallying cry to get more people engaged. No one expected it to become as big a splash as it did.

Tim Rayden is the forest restoration lead at the Wildlife Conservation Society, where I was employed for five years. He works closely with Trillion Trees and has spent the last few years developing new projects in forest protection and restoration. I called him up one day to talk about the costs and potential for scaling what Bethanie Walder at the Society for Ecological Restoration had referred to as five-star restoration: projects benefitting biodiversity, human health and well-being, and climate by fully recovering native ecosystems.*

* I later wrote a commentary about the real costs of planting trees for *Scientific American*, which was published in January 2021.

"We are never going to be able to grow a tree in the Bismarck Mountains in Papua New Guinea with seed collected from natural forests for less than a dollar per tree," Tim said. "If you want to scale up *and* prioritize areas of high biodiversity, the cost increases." When Tim was a colleague of mine in 2020, Trillion Trees had been trying to promote the idea that biodiversity is also important for harnessing the potential of trees in climate mitigation. A few years later, in 2023, a team of scientists presented scientific evidence showing that protecting and restoring wild animals and the roles they play in ecosystems can enhance natural carbon capture and storage. A diversity of animal species with medium to large bodies, for example, can aid seed dispersal that supports the germination of large-seeded trees with carbon-dense wood and reduce plant competition through herbivory, as well as enhance the soil nutrient supply and organic carbon storage. The issue is that when investors and governments focus only on the estimates for carbon sequestered as fast as possible at large scale, the economics don't favor the small-scale five-star efforts.[4]

To describe these projects, Tim used the word *boutique*:

Boutique (*noun*)
A small company that offers highly specialized services or products.
A business or establishment that is small and sophisticated or fashionable.

He wasn't implying that a company or business should run these projects; Tim chose the term to emphasize how much local context matters—that every project is different and highly specialized. Restoring a degraded natural forest in a hotspot for biodiversity requires very different work—often at smaller scale—than turning a cattle farm or other agricultural land back over to some definition of forest.

Tim said, "To me, *boutique* means not a small-scale project, but a project that doesn't have ready mechanisms so you can drive it to a landscape level. How do you make replicates of it?"

When I called him back to probe further, Tim opened a slideshow of a presentation on the project he had mentioned in Papua New Guinea. Sharing his screen, he scrolled past the Solomon Islands and Micronesia to land on the eastern side of Papua New Guinea. As he zoomed in, the steep contour lines across the terrain caught my attention.

"You can see that the islands are far out there, but the work is in the mountains—even more remote. There's no road access. They had to charter a plane and fly in materials that they walked in to start the nursery." He zoomed in further to five polygons where the planting would occur. "They're going to plant a mix of species in these deforested areas. Don't get me wrong; it's very cool work, and it's great to be supporting this. But even within each one of these five plots, every community wants to do its own thing. They might plant different species or species in different proportions. This is my extreme definition of boutique."

His argument was that there's a disconnect between what small-scale five-star restoration that promotes natural forests might require on the ground and what many people have been idealizing as large-scale tree plantings for a big chunk of the climate solution portfolio.

Tim offered another example from Nyungwe National Park in Rwanda, north of the Burundi border. Often the result of people using smoke to calm bees and collect honey, fires had destroyed thousands of forested hectares inside the park.[5]

"Bracken ferns colonize the burned areas. It creates an ecological barrier that you need to overcome to do any restoration on site. They slash down the bracken ferns and clear the dead organic matter that gets matted down below. That layer is suppressing the viable tree seed in the soil." Getting rid of the bracken enables regeneration to occur without planting. The seeds are still present. They need to be liberated so that the forest can return.

"This work is not the same everywhere," Tim added, "because everywhere the conditions have evolved to be different."

"Aren't there guiding principles?" my husband, Matt, asked me one evening when I was rambling about the range of approaches out there for increasing forest cover.

"Yes," I grumbled, a bit perturbed by the logic that *obviously* there should be a perfect prescription and a plan for widespread distribution.

Matt is trained as a doctor, an internist. After residency, he never practiced medicine. He always liked mathematics and problem solving by discerning patterns from a lot of information. So, he chose a path in public health instead of direct patient care, focusing on emergency responses to infectious disease outbreaks. It requires integrating medical knowledge with understanding of human behavior as well as local, national, and international governance issues to deliver on any large-scale effort that benefits many people. He has spent a lot of time in Africa, trying to halt the transmission of diseases like Ebola. But when the pandemic came to us all with COVID-19, work got, well, "very interesting" closer to home.

"The UN has ten principles for ecosystem restoration," I told him. "There are lots of other frameworks too." He caught my tone, skeptical of a one-size-fits-all approach to deliver millions or billions of trees, let alone a trillion.

"And they probably say some very general things like 'engage local people,'" he replied, getting practical.

"Exactly."

More credit is due to the United Nations than I was giving in the moment; the principles are well intentioned and thorough. For principle two on "broad engagement," for example, the opening reads, "All stakeholders, right-holders, and especially under-represented groups (e.g. local communities, Indigenous peoples, ethnic minorities, women, youth and LGBTIQ+ people), should be equitably and inclusively provided with opportunities to be engaged and integrated in meaningful, free and active ways. Such inclusive participation is necessary for achieving the desired outcomes of restoration over the long term, and should be promoted as much as possible throughout the process, from planning to monitoring." Other principles address benefits to nature and people, measurable goals, the causes of degradation, and more. *Forest landscape restoration*, a process promoted by the International Union for Conservation of Nature, highlights similar principles: focusing on landscapes, restoring ecological functionality, allowing for multiple benefits, recognizing a suite of interventions are possible, involving stakeholders, and tailoring to local conditions. But yikes, read either set, and they can come across as super vague in terms of what anyone involved should do in practice.[6]

Like anyone supporting the movement and hoping for beneficial outcomes, Matt was searching for some way to get from the case-by-case treatment to rapid implementation of global forest health measures. He faces similar challenges in his public health work, but he didn't have the answer either.

"It's an interesting contrast to global public health," Tim Rayden at Trillion Trees later commented when we revisited the tensions between boutique efforts and large-scale impact. "Hah!" he exclaimed, as he began playing with the analogy and its implications. "Imagine you are administering health care to the world's forests, and you realize every forest is different and so is how that forest ended up in whatever degraded state it's in. So, doc, how do you problem-solve that?"

As one global forest policy expert told me, "The private sector is finally starting to invest resources at a scale that could make a difference. But this unprecedented step has created a problem we weren't expecting to have: the challenge of using those resources to do the work effectively." He said that people want results fast. "It's like producing a vaccine or electrifying cars. Even when there's a solution in hand, it takes time for uptake in more communities and contexts."

I asked many of the people I interviewed about projects they'd worked on or come across one way or another that stood out as model efforts. In the years I spent following the news related to tree plantings, I too became accustomed to critical views and reports of failures in catchy headlines: "Most of 11m Trees Planted in Turkish Project 'May Be Dead,'" "'Greenwashing': Tree-Planting Schemes Are Just Creating Tree Cemeteries," "Tree Planting Is Booming: Here's How That Could Help, or Harm, the Planet." There were exposés I found or people sent me, like "The Fairy Tale Forest," which highlighted how the organization Plant-for-the-Planet "advertises money for climate protection with dubious promises and questionable figures from companies and private donors." (Finkbeiner, the founder, wrote that the allegations were "all unfounded.") That article ran in Germany. In October 2022, *Yale Environment 360* published "Phantom Forests: Why Ambitious Tree Planting Projects Are Failing," which identified all kinds of problems: wrong trees, wrong places, little or no follow-up monitoring, leakage (although the reporters didn't use that term), lack of follow-up care such as watering, competing demands for land, and "bad relations" between forest planters and locals.[7]

Even with all this critical coverage, I'd learned a lot of failures still go unreported. What organization, government, or funder wants to share how many trees didn't survive or expose any other unexpected complication that arose with more than just the trees?

Kevin O'Hara at 1t.org told me bluntly, "You can find the terrible examples in the media of, say, this monoculture plantation that displaced Indigenous people. The bad stories are easy to tell; they make the news." Other people I'd interviewed called it a click-bait narrative that all tree planting is bad. "But we don't have that same attention put toward the community that is thriving now or receiving additional income streams for their effort and care," Kevin continued. "It's harder to say, for example, 'Hey, look at how much progress we're making on reducing the heat island effect here in this one place.'"

It's simply easier to find something that went wrong at a moment in time than to track a project over years or decades and say, "Oh, hey look, news break, it worked!" But the good projects are out there too.

My question about favorite projects, or "pillars," of the movement first emerged on a bench in the Zurichberg forest before Tom Crowther and I wandered through the trails to his forest bathing spot. We were futzing with our smartphones in an attempt to open Restor—the sleek online data platform that was developed with the help of Google Creative Labs to be a hub for restoration and conservation sites around the world. That was early spring of 2022, and there were about 100,000 projects in the system. The platform had launched the previous year with about 50,000 initiatives on board. Not all the project locations and details were posted, given each submitter could opt in or out of sharing widely. But we could still take a quick tour of "all hands on deck" from tens of thousands of sites that were publicly available.

"It's called Desta's Coffee," Tom said, as he kept trying to load the data and scan through the many projects in Ethiopia. He was frustrated with the lag: "For some reason, my phone seems to be the only phone in the world where the Restor points don't load! The irony!" We laughed. "I wouldn't say Desta's coffee is a pillar as in it's better than everyone else, but I just know the project personally, and I love it." Desta Daniel Kebede, whose father is Ethiopian and his mother Swiss, has a large farm in Ethiopia that protects forest by planting native coffee trees in the sunny patches.

"The farmers around that land need irrigation and fertilizers to keep their crops, but Desta doesn't. The forest is doing its job."

I later downloaded Desta's project information from Restor when I had a better connection; I zoomed into the plot. From the satellite view, I could see what Tom had called the human footprint—a patchwork of farmlands in shades of brown where forests had been. Bordering the intensive farms was Desta's polygon outlined in white; inside were 108 hectares of verdant forest. What appeared as contours of broccoli from the bird's-eye view was clearly more ancient forest than farm.

"He sells their coffee here in Zurich. Now other farmers in the area are starting to do the same thing with their land." The company's website promotes coffee produced by "the most sustainable methods in the world" and details the direct role of farmers on site, intact "wild flora," and coexistence with wild animals. Desta and his partners were setting up beehives for honey too.

"They are economically incentivized to protect nature," Tom said. "That's the dream."

But Desta's Coffee is just one of thousands of projects. I imagined Tim would have called it boutique as well, given its size, the very specialized and localized incentives for protection and restoration, and the niche market for a product in Switzerland.

Are those trees sequestering carbon? Of course. Is anyone measuring the carbon on site? I honestly don't know from my research. Restor, however, provides an estimate of current sequestration and potential: considering aboveground biomass, Desta's Coffee was already harnessing 78 percent of the carbon potential on that land.

Pretty good! I thought, staring at the project's carbon summary online and wondering about the margin of error on the numbers too. *But not quite good enough to achieve all the carbon potential that everyone involved in this movement would like or is counting on.*

I selected other points for projects in random locations and clicked through the carbon data: 69 percent unrealized, 74 percent unrealized. I felt both hope and dismay in the same moment. *Maybe in time those efforts will realize more. There's so much potential,* I thought. I could also see the danger if governments or companies count on realizing more of the forest carbon potential as permission to continue emitting.[8]

I kept a file of favorites that emerged from the many treekeepers I came to know over the years of my research. Only when I reviewed all those projects together and compiled highlights could I internalize the diversity of endeavors. This is something new under the sun. This is all hands on deck for the load-bearing walls of this place we call home.

The monkey-ear or elephant-ear tree is named for the curious shape of its seed pods, which are round and textured around the edges, like the ear of an elephant or a chanterelle found in a forest. In its full grandeur, the crown spreads wide in a dome, a green umbrella resting upon the earth. In Panamá, these trees are known as corotús. The corotú is just one of nearly eighty species that community members are growing in a dozen nurseries across the Azuero Peninsula.

"They are the mothers of the forest. Get a big corotú and you have a complete ecological system," said Andrew Coates, a Brit who has been working on the peninsula for nearly two decades and now leads operations for a large-scale forest-restoration project in the region. The current land use is cattle grazing, but productivity has declined in recent years with more than a hectare of land required to support one cow. The goals: reforestation and natural regeneration on at least 10,000 hectares by 2027, an ecological corridor restored, rural livelihoods transformed with new income streams, and about 22,000 tonnes of carbon (81,500 tonnes of carbon dioxide) sequestered annually for three decades or more. That's equivalent to taking about 18,000 gas-guzzling cars off the road each year.

Robin Chazdon, who had criticized the Global Potential study early on, and Sarah Wilson, a tropical ecologist, conducted in-depth reviews of twenty projects that have been implemented. Those case studies are linked off Restor, and a number of people whom I interviewed mentioned them too. A few highlights from the flagship collection:

"In one decade, farmers in the Intag Valley in Ecuador went from clearing to conserving and restoring forests. . . . A local NGO, Defensa y Conservación Ecológica de Intag (DECOIN), helped people make the critical connection between healthy forests and clean, abundant water. . . . They helped thirty-eight communities establish small-scale watershed reserves, introducing communal land arrangements and the practice of forest restoration using native species."

"Vast areas of the Maradi and Zinder regions of Niger were transformed from severely degraded farmland to agroforests through farmer-managed natural regeneration (FMNR). . . . More than 90 percent of Maradi's population now encourages selective trees to grow on farms. FMNR transformed five million hectares from wasteland to agroforestry as individual farmers invested in trees."

"The Scottish Highlands have been largely devoid of trees for centuries. Forest clearing was followed by the eradication of large predators (wolves and lynx) and a dramatic increase in grazers, mainly deer and sheep. To date, high grazing pressure has prevented forests from growing back, and large hunting estates have prevented actions to control deer. The NGO Trees for Life (TFL) rewilded thousands of hectares to native Scottish forest." The project demonstrated that near remnant forests can regenerate naturally if grazing pressure is reduced. TFL also developed methods to propagate and plant native tree species.

Robin and her colleagues had gone back to interview the local people involved in each "case," which was extensively researched and independently verified. "We wanted to find projects that had been around for quite a long time," she told me. "You know, the real leaders that have withstood the test of time and have had long-term outcomes and impacts. It turns out most of those are community-based projects because, well, the more top-down ones tend to be short term." Local people may not be empowered to keep projects going in the top-down model.

Each project wasn't just about planting trees but instead was tailored to who and what could protect and grow trees—culturally, politically, economically, ecologically—in place. Reading through the files, I got the feeling that these were the hidden stories of the complexities behind the simple big idea for forest recovery across the world.

Trillion Trees offered a favorite example in Central America—a project called Selva Maya that spans Guatemala, Belize, and Mexico. It began with recovering degraded forest inside Guatemala's Maya Biosphere Reserve, which was designated in 1990. Then land had become occupied illegally by cattle ranchers and drug traffickers. The Selva Maya initiative brought new sustainable jobs to remove pasture grasses and bring the forests back. For the first time in the reserve's history, forest cover began increasing on its land. I watched a video on Twitter from a trap camera there: a tapir and its baby, wandering through a young forest—a symbol, if you will, of the habitat re-created.

Contacts at The Nature Conservancy pointed me to the Xingu Seed Network, an Indigenous-led effort to collect and sell seed from more than 200 different species for restoration in the Amazon and Cerrado, tropical savanna located in central Brazil. The network brings together women from about twenty-five different communities to supply native seed for widespread use.[9]

The favorites flowed from Kevin O'Hara at 1t.org. "The revitalization of thorn forests in the Lower Rio Grande Valley comes to mind," he said. "Gisel

Garza is a seed collector; she's with American Forests. She goes from tree to tree in this phenomenal thorn forest on the border between the US and Mexico." Amidst the rugged-looking trees, there are also more than 1,000 species of plants, 500 species of birds, and some 300 species of butterfly, in addition to the only US population of ocelot, an endangered wild cat with a spotted coat. "It's not even a forest in the way you think of a forest; it's shrub, really dense and low-hanging thorn forest," he explained. Gisel and other partners have been boosting collection of native seed that will be more resilient to future climate change. Plot by plot, the team is rebuilding habitat that's been decimated by urban development and stressed further by climate change. They're prioritizing drought-tolerant species. They've estimated the carbon potential. They're aiming to sustain the diversity of species.

He pointed to another effort in California in response to the Camp Fire in 2018, the most destructive and costly fire in the state's history. "They're thinking intentionally about what it means to restore an area decimated by historic fire. What do we need to be doing on thinning? What are we doing on replanting? Are we using technology? Do we replant? Indigenous people have used fire to manage forests for millennia, but when the fires burn so hot now, they're burning down the topsoil. So, natural regeneration isn't possible." There are all kinds of strategies for assisting that regeneration. Island planting is one; the idea is to plant small islands across large landscapes to create long-term seed sources for the forests to expand again.[10]

The Great Green Wall, launched by the African Union in 2007, was Kevin's favorite international example because the scale of the effort is astounding, and it's trying to solve multiple problems at once. Eleven countries have joined the effort to combat land degradation and restore native plant life, particularly native species, across 4,831 miles—from one side of the continent to the other. Kevin's colleague told me, "If you get this right, you will stop civil wars. You will stop famines. You will stop drought. You will stop species from going extinct."[11]

None of these favorites that people described to me were monoculture plantations. Matt Hansen, the remote-sensing scientist who helped revolutionize global forest monitoring, told me that Uruguay has had the greatest increase in tree cover in his records. "It's all eucalypts and some pine, though. The country has taken traditional pastureland and turned it into forestry, or soybean farms." Although a satellite sees more tree cover, what the trees comprise together as "forest" has changed dramatically. Plantations never came up in my interviews as models for the kind of work that more people should be doing.

The UN Decade on Ecosystem Restoration has embraced the whole suite of approaches—different kinds of efforts across the restoration continuum—and I think that's what Kevin means by "all hands on deck" for 1t.org too. It's not only all hands on deck for five-star forest ecosystem restoration. If you're financing a forest restoration and you're single-mindedly focused on the return on investment with respect to carbon sequestration, the Bismarck Mountains or the thorn forests on the US–Mexican border are probably not where you'd go first. Yet there are lots of reasons to restore forests in these areas, including the carbon sequestration service.

"I don't think there's been any specific policy about the approaches included in the UN Decade or not," Robin said when I probed about what counts in the global push. She's been an advisor to the UN Decade's Task Force on Best Practices. "Take the National Regreening Project of the Philippines. It's something like eleven million hectares, and most of that is monoculture plantations. So, nobody is bound to anything. You can do whatever you want." Canada had recently made another commitment to planting, and that would include some monoculture.* "I mean, that's what they do in Canada, right?"[12]

I lost track of how many people pointed me toward the International Small Group and Tree Planting Program (TIST). Active in Kenya, Tanzania, Uganda, and India, it's been around since 2004. TIST hits scale through the sheer number of local farmers involved in planting trees on their lands—thousands of villages and nearly 200,000 farmers to date. TIST

* Plans also include planting fire-resistant deciduous trees near communities, instead of relying on more flammable conifers.

calls its work "time-tested forestation." I watched an online time series of each program in expansion; more and more yellow dots popped up for reversals underway. They covered degraded lands surrounding Mount Kenya and in southwestern Uganda, as well as in other areas. I scrolled through pages of tree audits from TIST's rigorous and transparent monitoring system. From the size of the trees reported over time, I could see the forests growing.[13]

Farmers collect the seeds, plant trees, and maintain groves; they're economically incentivized, as Tom Crowther would say, by fruit and nut trees and the sale of other nontimber products. Verified TIST results show that current farmers have received about $140 million in benefits—$8 per tree planted "and kept alive."[14]

Those estimates don't even include the returns to each farmer from selling the amount of carbon sequestered by the trees as an offset for emissions made elsewhere on the planet, another mechanism for marrying the ecology and the economy.

"Why would companies be interested and involved in this?" I have thought about that question, which was posed at a 1t.org press conference in Davos, Switzerland, for quite some time. I've asked the same of representatives at various companies and organizations too.

"Marc Benioff at Salesforce highlighted net zero and the global climate crisis," I recalled to Kevin. "While his response was moving, and it captures some of the motivation, I'm thinking it doesn't capture it all. What's your answer to the question?"

"When all this started, carbon was the focal point," he replied. "But even Salesforce is saying net zero and 'nature positive' now." For net zero, companies leverage forest projects and other strategies for sequestering more carbon to reduce their emissions or remove them from the atmosphere over time. Nature positive adds more layers to the motivation for better stewardship. Kevin referenced a report that the World Economic Forum had published in 2020, stating that over half of the global gross domestic product, $44 trillion, is threatened by the loss and degradation of nature because human societies and economic activities fundamentally rely on biodiversity. "There's a growing recognition of the interconnectedness between economic success and nature, of how much we depend on nature for clean water, for clean air, and carbon"—and for so much more, like food production and public health too. He said that more companies are considering, "If nature collapses in this place, what impact does that have on our bottom line for *x*, *y*, and *z* reasons?"[15]

Kevin's response brought me back to the classroom at Brown University, when I was nineteen or twenty and first learning about environmental problems as externalities, the consequences not accounted for in the cost of producing a good or service. At the time, we were reading a seminal study that reported the economic value of services coming from nature, such as water supply, nutrient cycling, soil formation, pollination, climate regulation, and more. Robert Costanza and his coauthors estimated $16 trillion to $54 trillion per year when the study was published in 1997. That study was among the first to put an economic value on the ecosystem services and the world's "natural capital"—the soil, air, water, and other assets that nature has to offer. Putting natural capital into practice is an effort to assign value to the services that nature provides and thus account for them in human activities across the world. If we can assign value to a service that nature delivers, such as nutrient cycling, water supply, or pollination, then payments to landowners can incentivize better stewardship.[16]

"I think what we're seeing now is a growing recognition of what had been viewed as sort of fringe or far left," Kevin told me. "Nature is being talked about at GreenBiz and in boardrooms in ways that it only was in academia years ago."

Even with my youthful idealism as an undergraduate student, I thought that the possibility of people and governments consistently accounting for natural capital, our world's stocks of natural assets such as trees, minerals, or the atmosphere, seemed far off, if not far-fetched. It also felt a bit sad to me that money might be the only way left to reawaken the desire to protect and defend nature, which I would come to see as innate much later through my son.

Applying this idea of natural capital would require modern society to figure out how to value nature and its abundant services in a capitalist system. Such a shift could help capture the costs of producing goods and services, thus reducing negative impacts to the environment that are so often overlooked. It could incentivize a rebirth of care and stewardship.

Instead of ignoring the degradation and leaving it for others to repair or endure its consequences, can we ever create a new normal where prioritizing and investing in nature is the standard? Efforts to support this kind of change at scale are underway.

PART III

AT SCALE

In a field
I am the absence
of field.
This is
always the case.
Wherever I am
I am what is missing.

When I walk
I part the air
and always
the air moves in
to fill the spaces
where my body's been.

We all have reasons
for moving.
I move so
to keep things whole.

—Mark Strand, "Keeping Things Whole"

12

INCENTIVIZE

There are a couple stories about what did and did not happen on May 9, 1997, and I can't help but wonder what might have unfolded for the world's forests had things gone differently that day. In the version covered by the media, President Bill Clinton ventured to the famed Braulio Carrillo National Park, a tropical forest that is home to thousands of plant species, hundreds of bird species, and a diversity of amphibians, reptiles, and mammals, including jaguars, pumas, and peccaries. The park is just outside San José, Costa Rica's capital city. The day's focus, as CNN reported, was on environmental protection and the need to ramp up efforts in Latin America, given the pressures of population growth and deforestation. Having reversed its trajectory of forest-cover loss to forest-cover gain and effectively turned the results of its treekeeping into more tourism, Costa Rica had become a shining example of successful conservation and reforestation.[1]

I'd learned about the events of that day in my effort to track down some of the earliest examples of *carbon offsets*: reductions or removals of carbon emissions that make up for emissions that occur elsewhere. Today, among scientists like my former advisor Eric Lambin and others who study forest transitions, Costa Rica is widely recognized as a pioneer in implementing payments for ecosystem services, incentives offered to landowners in exchange for managing their land to provide some sort of ecological benefit. Payments to avoid forest degradation or reduce emissions through carbon credits and offsets later became financial mechanisms used across the world to facilitate better stewardship of existing forests and also to grow new ones. So, I wanted to go back to the people who were involved in some of the earliest approaches to paying for carbon sequestration through trees and other efforts to offset emissions.[2]

A downpour hit hard during Clinton's visit to the park. "You were very, very emphatic about wanting to visit a rainforest," Costa Rican president José María Figueres joked with him. "We have some thunder and lightning on order in a few minutes."

Before the paper got too wet, Clinton encouraged US Interior Secretary Bruce Babbitt to sign an agreement regarding a few initiatives on climate and the environment. According to the terms, the United States would assist Costa Rica in finding ways to put more electric cars, buses, and motorcycles to use to reduce smog. A US company had already brought one electric bus to the country, and others would follow. Costa Rica would provide the United States with computer software for tracking biodiversity in its rainforests, which Americans wanted to use in their national parks.

In a probing question raised at the press briefing that same day in San José, someone asked Babbitt how he felt about the new commodities, "which are essentially like emissions credits," that Costa Rica had begun trading or selling across international borders. The inquiry was about commodifying forests through carbon credits and emissions reductions achieved by nature, which people would later call natural climate solutions. Generated by forests in Costa Rica, these first credits also laid the foundation for what would later make forest restoration possible through for-profit businesses, like Terraformation. They expanded the ways to finance restoration and reduce forest degradation beyond the more traditional sources of funding through governments, grants, or donor dollars.

A credit for carbon can be certified by governments or an independent certification body, and it represents an emission *reduction* of one tonne of CO_2, or an equivalent amount of other greenhouse gases, such as methane. Also known as a carbon allowance, it's like a permission slip for emissions. A company gains permission to generate one tonne of CO_2 when it buys a credit, and that credit is also tradeable. A carbon offset typically refers to a *removal* of greenhouse gas emissions, such as trees to sequester carbon, or *avoided* emissions achieved through renewable energies or avoided deforestation, for example; the offset compensates for emissions occurring elsewhere. People today use the terms *credit* and *offset* interchangeably, so it's also helpful to think of "the claim" made when distinguishing the two. The purchaser of a credit can "retire" it to claim the underlying reduction (or removal) and apply it toward their own voluntary emissions-reduction targets or a mandatory reduction set by a government. "It's like saying, 'By retiring this credit, I have offset my own emissions,'" one expert explained to me. That claim transforms the credit into an offset. In its compliance program, the state of California even uses the terms in combination, referring to "offset credits" for credits that could be used as offsets. The possibility of great inequity between landowners who maintain the forests and investors who may profit by owning the credits is one of the many critiques of carbon trading. It surfaces in compliance carbon markets, through which entities obtain credits and surrender offsets to meet regulated targets like in California, and the voluntary carbon market, the current system for voluntarily trading reduced, removed, or avoided emissions. More on all this in due course.

Just months before Clinton's visit, Norway had purchased 200,000 tonnes of mitigated carbon dioxide from Costa Rica for $2 million in the first certified carbon transaction in the world. The funding went toward conservation of existing primary and secondary forests, along with reforestation of lands used unsustainably for livestock. Norway gained the ability to claim it had reduced its own emissions by facilitating forest carbon sequestration in another country; Costa Rica could use the money for the work required to deliver the service.[3]

At the press briefing in San José, Secretary Babbitt offered a diplomatic answer to the question about the new commodities. His response was indicative of the good intent of this clever strategy to pay people in one place for protecting and restoring the carbon-sequestering capacity of their forests to benefit many more people and the climate system. "It's a very important idea that works to the benefit of developing countries and the United States," he replied. "It is a market mechanism which acknowledges that the output of carbon dioxide into the atmosphere is a global problem. A unit of carbon dioxide emitted in Costa Rica has as much effect on the climate in Arizona, where I'm from, as it does in San José, Costa Rica."

He called carbon dioxide globally fungible, meaning what's emitted in one place is comparable with what's emitted in another. Similarly, releasing a unit of carbon dioxide in one place can be offset, or balanced, by drawing one down from the atmosphere elsewhere. He added, "Economics would say, 'You ought to try to get the most bang for the buck.'" Secretary Babbitt was suggesting that if it's more expensive for a country to reduce its own emissions than to pay another one to do the work, it saves money to go elsewhere and purchase the service—through trees or other mechanisms.

In the other story of what *didn't* happen during Clinton's visit to Braulio Carrillo National Park, there was a planned exchange between the two presidents. As the Costa Rican ministers who were present that day shared with me, President Clinton would have handed President Figueres a twenty-dollar bill while standing against the backdrop of that magnificent cloud forest. Figueres would have passed back a piece of paper for an offset credit, verifying one tonne of CO_2 mitigated.

"The press team at the White House pulled it before it happened," Alvaro Umaña, Costa Rica's first minister of the environment (1986–1990) under President Oscar Arias, told me when we met on the Stanford campus. The White House lawyers had advised the press office to change the plan. I'd already interviewed Alvaro remotely about Costa Rica's reversal in forest cover and ways of incentivizing forest protection and restoration. Yet this anecdote hadn't surfaced until we talked more about climate policy and missed opportunities of the past. Alvaro said that he and his Costa Rican colleagues had printed the certificate in preparation for the US president's visit.

"It got pulled because it would have set a price for carbon?" I asked him in shock. I was astounded that they had almost settled on an amount that seemed quite high, even before considering inflation. In 1997, $20 would be equivalent to nearly $40 in 2023. For contrast, the International Monetary Fund reported a global average of carbon pricing in 2021 at $3 per tonne, far below what's necessary to establish and sustain a project that removes carbon, avoids leakage, and endures over time. (Norway had paid Costa Rica about $10 per tonne in the first certified carbon transaction, and the $2 million total was less than half the actual project cost at that time.) A more recent article in the *Financial Times* showed a peak in pricing for nature-based credits at around $15 per tonne in January 2022, falling steadily to around $5 per tonne a year later.[4]

"Yes, it would have sent a signal on pricing," Alvaro agreed. "It would have also showed the US was really moving on addressing climate change, but potentially people might have perceived the exchange as the US was buying up forests in Costa Rica."*

My stomach churned, thinking about the lost opportunity to value carbon at such a price so many years ago. It would have established a higher economic value for trees. It might have influenced the price of a tax per tonne of CO_2 emitted, for example, had US policymakers implemented a nationwide carbon tax. It could have helped incentivize other reductions because if it's not actually cheaper to buy the carbon units that are mitigated elsewhere, a company or country would make more changes to its own practices. It's like the choice of riding a bike or driving a car; if the price of gas is higher, you might ride your bike more often.

From 2020 to 2022, I'd worked on a report with Trillion Trees about estimating the real costs of restoring forests. In some ways, the research was motivated by the widespread misperception that planting a tree and, by association, growing a forest over decades could cost $1 per tree, or even less. Many organizations, such as One Tree Planted, the National Forest Foundation, Grow Clean Air, and ReGreen, were effectively attracting supporters with the promise of planting one tree for every dollar donated. Felix Finkbeiner's Plant-for-the-Planet was offering €1 trees. Eden Reforestation Projects was going as low as fifteen cents.[5]

As our report stated, when considering the average carbon-sequestration rate of recovering natural forests, one tree sequesters about 200 kilograms, or 0.2 tonnes, of CO_2 over twenty years. The problem is that if people look at a market price of $5 per tonne of CO_2 and do the math, they end up thinking, "Great! Well, that should give me a tree for every dollar." The market value of the carbon dioxide, as well as the marketing

* I contacted President Clinton's office and his former White House press secretary to confirm the exchange and get another take on its implications. Despite my numerous inquiries, neither responded.

appeal for garnering more support, may be driving the $1-per-tree valuation instead of rigorous consideration of what growing that tree requires over time. I'm not an economist, but I understand market dynamics as generally being very effective at driving competition and reducing the cost of accomplishing an objective. So, in this case, what if the market pricing falls short of the actual project cost?[6]

A figure of $5 per tonne is far below most estimates of what's adequate to generate emissions reductions from forest-based approaches. A 2020 study, for example, estimated the cost of carbon sequestration in recovering secondary forests of Brazil's Atlantic Forest at $66 per tonne. A 2018 study that used data from Cambodia found that just keeping a forest standing required $30 to $51 per tonne of carbon dioxide to break even on costs for not using the land for other purposes, verifying and certifying the carbon, and ensuring long-term stewardship. In a book I'd read by John Doerr titled *Speed and Scale: An Action Plan for Solving Our Climate Crisis Now,* Christina Figueres, the former executive director of the United Nations Framework Convention on Climate Change, called the current (2021) prices "ridiculously low" at "generally between two and ten dollars" per tonne. She stated that to make a real difference, the price would need to go up to $100.[7]

When I spoke with René Castro Salazar, the Costa Rican minister of the environment and energy who was also there that day in 1997 and could confirm the details, he seemed distracted at first, which threw me off. I tried not to get rattled and began asking about Clinton's visit. I needed another behind-the-scenes perspective, and I'd contacted him to confirm the details.

"Check your email," he said, as his attention returned to me on the screen. He'd been searching for something to send me. I toggled between windows.

The subject was "CO2." Inside was an image of a certificate from 1997 that read "Greenhouse Gas Emissions Mitigation Certificate."

"Whaaaat?" I exclaimed. "There it is!" The paper was light green with darker-green writing. It was difficult to read the old and fuzzy scan. I squinted to make out the small text:

> The Costa Rican Office for Joint Implementation of the Ministry of the Environment and Energy of the Republic of Costa Rica certi-

> fies that the bearer of this official certificate has contributed to the improvement of global climate by providing additional financial support to specific national sustainable development projects in Costa Rica that have mitigated a quantity of greenhouse gases in carbon dioxide (CO_2) equivalent units of ________ metric tons.

The quantity had been filled out in black lettering: 1,000. The certificate was an unsigned sample, but a note at the bottom referenced that first transaction in the world, which had occurred with Norway prior to Clinton's visit. Alvaro has since kept a similar credit with "1" typed into the blank for the $20 exchange that never happened.

Minister Castro sent me another email; this one included a copy of the first signed certificate, which was in Spanish. There was a photograph of him with the certificate in one hand and a check in the other. He was standing with the representatives from Norway who had facilitated the transaction. He told me the Costa Ricans had set the price too low by focusing on the opportunity cost of their land; they'd failed to consider what it would cost for Norway to reduce its emissions via other mechanisms.

Carbon also wasn't the full story for Costa Rica. It was offering what Alvaro called "boutique carbon credits," given that forest protection and restoration could also maximize biodiversity: "Our carbon is like a fine wine, with a bouquet of water and more than a hint of biodiversity." It had traded relatively low-cost carbon reductions through land management. A higher price could have done even more for the world's forests and all the life connected to them, as the offset strategy later spread.

Without an adequate price set for carbon when the offsetting began, a race to the bottom would emerge through market forces. If one tonne of carbon emitted is the same as one tonne sequestered somewhere else in the world, then anyone looking to reduce emissions can seek the cheapest pathway to achieve their goals. But when it comes to growing trees over time and restoring forest ecosystems, the cheapest way won't do the job. A system that economically incentivizes people to plant trees for one benefit isn't enough; it needs to incentivize keeping them and the forests they become.

I felt a mix of emotions as I stared at the photographs, these relics of an actual exchange for offsetting emissions between countries and an exchange that never happened. Alvaro had also sent me a photo of him with Presidents Clinton and Figueres. Clinton holds a frame with a blue morpho butterfly mounted beneath the glass. The three men are each wearing a rain jacket or poncho, and they're glowing in celebration. In the photo of the Norway exchange, Minister Castro's subtle expression appears to be one of pride as he holds a certificate and a check together. I felt the optimism and hope that they must have all had nearly thirty years ago. I felt frustration and dismay for the consequences that arose later—the false assumptions that offsetting could resolve the climate crisis and that planting trees could be an easy fix for our climate in general.

The great passage of time also disappointed me. Those photographs were a testament to how long people had been innovating to address the climate crisis without any one solution taking hold to do the job.

I needed to go back even earlier to find the birth of the offset idea and understand the early intentions. Even though it took decades to gain global traction, the plan to pay for forest carbon sequestration had certainly helped ignite this something new under the sun.

The *New Yorker* ran a cartoon in 1989 of a guru sitting cross-legged on a large pillow while men in suits stand in line, waiting to meet him. One businessman is turned to another and appears to be explaining something with excitement. The caption reads, "It's great! You just tell him how much pollution your company is responsible for and he tells you how many trees you have to plant to atone for it."[8]

When I first came across the cartoon, I couldn't believe it had been published that long ago. It seemed apropos of the media flurry surrounding the Global Potential study and the booming interest in tree planting that came decades later. ("We Can't Just Plant Billions of Trees to Stop Climate Change," *Discover Magazine*, July 10, 2019, or "Massive Reforestation Could Greatly Slow Global Warming," *Scientific American*, July 4, 2019, among many, many others.)[9]

"We got poked fun of even back then," Sheryl Sturges told me over a call. I'd come across her name in a short NPR story from 2015 about whether carbon offsets work. The reporter had called her "the inventor of carbon offsets," a bold claim but one indicative of her role in putting an idea into practice.*[10]

I reached out to inquire about Sheryl with Gretchen Daily, another mentor at Stanford, who is widely recognized as one of the pioneers of ecosystem services and the field of natural capital, the economic valuation of natural stocks such as trees or minerals. I figured Gretchen would know the history of early efforts to offset emissions before Costa Rica made that first carbon trade in 1997. She connected me with Roger Sant, cofounder of the AES Corporation, the global energy company that Sheryl Sturges was working for when the question of how to address the greenhouse gas emissions from power plants arose.[11]

Roger had started the company in 1981, and from the beginning, he aimed to supply energy by not only meeting the US Clean Air Act but "doing better." He was also on the board of the then fledgling World Resources Institute (WRI), which would later become a renowned research-based organization that aims to meet people's essential needs, protect and restore nature, and stabilize climate. Decades later in 2014, for example, WRI would launch Global Forest Watch, the online platform that still offers the best publicly available data on the state of the world's forests. Today, anyone can click anywhere on the globe to track the latest on both tree cover loss and gain. With the world in full view, it's pretty much impossible to see any blue pixels that indicate forest gain amidst all the red loss. But zoom in close, and you can see polygons of emerging forest sprinkled like fairy dust across our plant-covered continents.[12]

Roger was in a boardroom in 1986 with the founder of WRI, who was talking about climate change. "It was one of those moments in your life when you realize, 'This is a big deal,'" Roger told me over Zoom.

"I thought, 'We're part of the problem.' Here we are a little company, but we're building plants that are adding to the problem. Here we are

* The British American physicist and mathematician Freeman Dyson had published a paper in 1977 that first proposed controlling carbon emissions through tree planting. "Suppose that with the rising level of CO_2, we run into an acute ecological disaster," it posited. "Would it then be possible for us to halt or reverse the rise in CO_2 within a few years by means less drastic than the shutdown of industrial civilization?" Dyson ran the earliest estimates on the size and cost of a tree-planting program. (Dyson, Freeman J. "Can We Control the Carbon Dioxide in the Atmosphere?" *Energy* 2, no. 3 [September 1, 1977]: 287–91. https://doi.org/10.1016/0360-5442(77)90033-0.)

hoping we can make the environment better, but we are missing a whole class of pollutants that we'd never focused on.'" Feeling distraught, he didn't know what to do. He gathered his team at AES and told them what was bothering him.

"We've got to find a way to be a part of the solution, or not do what we're doing." He didn't want to abandon the business, which was just gaining steam and beginning to have an impact in the energy industry. Roger identified a coal-fired power plant the company was building in Connecticut; he was concerned it would contribute to the problem.

Roger wanted a solution, and Sheryl Sturges was the staff member who came back with an idea. Sheryl had come to AES with research experience in industrial energy use in the United States and the cost-effectiveness of conserving energy versus constructing nuclear power plants. In her early days with the company, she worked on ways to conserve energy to reduce dependency on foreign oil and approaches for making energy production more efficient. Roger thought there might be a technological solution to the CO_2 issue; perhaps the waste could be used in the enhanced oil-recovery process. Sheryl had read a lot about environmental issues, and she knew that deforestation was contributing to the climate problem. "So, I thought, 'Let's look at that,'" she told me.

"My sister worked at the Library of Congress so she was able to send me books and other resources," Sheryl recounted. "I came across one paper that said if you planted the Big Island of Hawai'i with *Acacia* trees, it would absorb the excess global carbon emissions at that time." She couldn't recall the study, and I couldn't find it either. That didn't matter; I knew the calculations and result wouldn't hold anymore. But the idea had stuck.

"I just needed to know that for our little power plant, I didn't have to plant the whole world! My scale was going to be so much smaller than the Big Island of Hawai'i. So, I felt like maybe we could actually do it—or some version of this. We didn't have connections to foresters, so I was trying to figure out how many trees we'd need. There were some basic numbers I could find about how much one tree absorbs. But then, what species would you use? Where would we plant?" She thought maybe they could reforest in areas with coal-fired power plants, then learned that any emissions released would be circulating in the atmosphere within a couple weeks.

Planting anywhere was fair game. Her story unfolded like a precursor of Yishan Wong's back-of-the-envelope calculations for a trillion trees.

Sheryl came back to Roger and said, "Well, here's a crazy idea. Trees sequester carbon. Why couldn't we plant some trees?" They didn't know how many trees that might require. About a week later, she came back again and told him, "Well, it's about fifty million trees." The estimate turned out to be high; it would be more like eighteen million today, given advances in estimation methods. Fifty million trees seemed like a lot of trees to Roger for offsetting the emissions from one power plant, but he was willing to give it a go.[13]

There was also evidence from strategies to address other pollutants that showed an offset system could work. In the 1980s, California had created an offset market for sulfur oxides and nitrogen oxides (SO_x and NO_x, or what sounds like "socks and knocks" among experts who enjoy a little word play). Massachusetts lieutenant governor John Kerry helped put together a response to emissions from coal plants that were contributing to acid rain. The program later became a model for the world's first large-scale pollutant cap-and-trade system, and Kerry helped develop the regulatory approach when he was in the Senate. The goal of the Acid Rain Program under the 1990 Clean Air Act Amendments was to reduce annual sulfur dioxide emissions in the United States by "ten million tons from 1980 emission levels."* Coal-burning power plants could buy and sell emissions permits, which were limited in quantity by policy regulation.[14]

The value of those permits rose incredibly high and then effectively went to zero, Roger recalled, still holding a sense of amazement decades later. "People were so creative that they figured out how to stop emitting sulfur dioxide," he told me.

"I can see how that would make for the perfect mindset," I said to Roger, as we talked about the SO_x and NO_x success. "As in, 'Okay, here's a new pollutant problem. Why wouldn't the same thing work?'" The key, I later learned, was sourcing low-sulfur coal from the West and shipping it

* US federal agencies have employed a mix of measurement systems since the 1970s. In 1991, an executive order under the Bush administration aimed to resolve the inconsistencies by requiring usage of the metric system. A language problem has persisted, however, given that some uses of *ton* still refer to metric tons (tonnes) without clearly indicating as much.

for use—another example of how solving one problem might still exacerbate another.

At the time of the tree-planting decision, Roger wasn't thinking about a global cap-and-trade system for carbon. He wanted to feel that AES was doing its best to reduce or eliminate its own impact.

Paul Faeth, who was trained in international development and had experience in agroforestry, ultimately led the team that designed the first offset at WRI. "AES asked us to find a project," Paul recalled, noting that they hadn't used the term *carbon sequestration* in the early conversations. They called it *carbon fixation*, and WRI convened a group of climate modelers and experts to discuss if using trees was even a legitimate strategy to reduce what was accumulating in the atmosphere. "Very quickly, the answer was yes." Among the AES criteria was that the selected project would be a "no regret" initiative, meaning it had to have additional benefits like supporting farmers through agroforestry activities or preserving biodiversity in case the mounting climate concerns didn't play out.

Paul and his colleagues at WRI didn't do a big tender process; they searched through existing networks and landed on a project in Guatemala. The project proposal was from a humanitarian organization called CARE, and it was best on cost. However, WRI also had other reasons to pick that one. Since the mid-1970s, CARE had been running a program called Mi Cuenca ("my watershed" or "my basin") to help farmers improve their water and recover topsoil in Guatemala. Activities included planting trees on steep slopes to capture water runoff and create natural terraces. It had delivered benefits to farmers and the environment, but it was still struggling on long-term funding.[15]

"It was a trustworthy organization," Paul told me, and investing in a project with a proven track record was attractive.

Mark Trexler, an associate in the Climate, Energy, and Pollution Program at WRI at the time, developed the methodology for quantifying the offset. In our correspondence, Mark told me that the first mitigation proposal he saw involved a complex plan to reforest parts of Australia, populating them with African endangered species, irrigating them with water from a massive desalination project, and using the salt to pickle the trees before burying them forever. Burying the trees could preserve the wood and lock up its carbon, as in the Carboniferous forests of the past. "In several ways," he noted, "it was prescient of where some of the conversations are today."[16]

AES put forward $2 million in funding, an amount that exceeded the company's net annual income in both 1987 and 1988. Leveraging that support helped raise millions more from other sources. It planned for fifty-two million trees in Guatemala over ten years to help 40,000 farmers. Activities included establishing agroforestry systems, planting community woodlots and reforesting, fostering natural regeneration of native forests, and reducing burning in existing forests. Mi Cuenca was rechristened Mi Bosque—"my forest."[17]

As AES developed its program, Sheryl hoped it would inspire other people to be "wildly creative in reducing greenhouse gas emissions" as cost-effectively as possible. Sheryl was inspired by California's SO_x and NO_x response; she dreamed that someday an international system of greenhouse gas offsets could exist.

The CARE project became known as the world's first carbon offset, but it had issues, as one might expect. It wasn't "certified," meaning a third party wasn't evaluating the effort and verifying that it had achieved its carbon sequestration and other goals. It took place on private lands, so CARE would offer trees and training to farmers who wanted to diversify their land use or reduce erosion. The agroforestry approach could increase maize yields, because the trees planted for erosion protection also fixed nitrogen and provided organic matter to retain moisture. Of course, over time it would sequester carbon too. But there weren't signed agreements with the landowners, raising a common issue that surfaces today with projects that receive funding from a few sources: Who owns the right to the credits? If two companies and a government invest in a project that generates credits, for example, it can take some careful accounting to make sure that multiple funders aren't claiming the same trees for their emissions reductions.[18]

"Who will follow up that those trees are growing?" asked former environment minister Alvaro Umaña when we discussed the challenges the CARE project faced. "How do we know that they are still there?" He said that he would bet very few of them are still there today. "The project in Guatemala brought a lot of lessons to the world about carbon offsets."

Later, when I went back to Paul Faeth, who had helped select the project for AES, and shared those reflections over email, he highlighted the text "very few of them are still there today." He responded, "Not those particular ones, no, but I expect that farmers are still planting fruit trees and doing agroforestry. The real benefit may have been in helping the farmers move away from unproductive slash-and-burn agriculture." It hit the mark for "no regret" with development benefits. Paul noted the higher yields, fruit production, enhanced income, and even improved nutrition that came from the effort.

An external assessment of the AES/CARE project's first ten years revealed that far less than the estimated carbon had been offset, due in part to the smaller total area of planting than expected. There were other problems, such as the increased conflict between landowners and local authorities as land use changed. Internal CARE memos suggested that the amount of carbon sequestered was unknown, given uncertainties in the modeling approach and missing data on deforestation and natural regeneration. I found media clips online that further mischaracterized the effort and its shortcomings on actual carbon sequestered as "allowing emissions in the United States to increase." Roger and Paul both told me, in retrospect, that the project might still have offset (or "balanced") the power plant, given that the original estimate was an overshoot and the plant closed early. But rigorous monitoring and reporting were lacking, so there is probably no way of knowing today.[19]

Another early AES project targeted reductions in methane gas that was generated through dairy production in India. The company also helped establish the Mbaracayu Natural Forest Reserve, which became the largest continuous remnant of the Interior Atlantic Forest in Paraguay. "Now the soybeans go right up to the park boundary," Paul told me. Sheryl showed me a photograph of her hugging a colossal *Ibirá-pitál* tree (*Peltophorum dubium*) inside the reserve. She called it a "grandmother tree" and said that without the protective effort, it could have been burned in land clearing. "There's a nature preserve and watershed protection and all sorts of benefits provided by those forests, aside from the carbon sequestration that the national park provides. It's because of the money that AES put up to balance another coal-fired power plant. You can see it; there's the forest, and there are the soybeans. It's very, very clear the project was successful."* The practice of offsetting had taken hold, and it wasn't only through planting.

I asked Sheryl what it was like to observe, years later, what the demand for credits and these natural climate solutions has become.

"Imagine you're sixty-five now," she replied. "You had this crazy idea long ago, when you were in your kitchen and your baby was sleeping, and you were writing proposals about how this offset market might work. But, oh my God, it actually happened!"

She said she was both encouraged and terribly saddened. "There are actual job titles in carbon accounting now!" There are also offset managers for businesses, third-party verifiers, and certification schemes with standards for monitoring and reporting. There is a voluntary market, where emitters can opt in to offset, and mandatory compliance markets for certain sectors in the United Kingdom, the European Union, and the state of California.

"We're also not as far along as I hoped we would be," she said.

An international mandatory system—the global cap-and-trade version for carbon of what happened with SO_X and NO_X—never came to pass. The Kyoto Protocol, an international treaty adopted in 1997, would have made strides in that direction by committing industrialized countries and economies in transition to limit emissions and allowing them to offset under what was called the Clean Development Mechanism. President Clinton signed the treaty, but the US Senate refused to ratify it,

* Paul's comments also point to persistent debates surrounding geographic leakage with offsets in general. Does protecting a forest in one area just lead to more deforestation outside the boundaries? Some experts say that continued land-use change in adjacent areas shows the protective efforts were necessary and effective. Others contend that a protected area may have caused increased deforestation elsewhere, thus negating the protective effort with losses elsewhere. This question as to whether a project has a net positive or negative outcome over time and geographic space has been raised in recent investigations of offsets that are based upon avoided deforestation and degradation (meaning, protecting a standing forest under pressure for conversion), as opposed to those generated by reforestation. (See, e.g., Greenfield, Patrick. "Revealed: More than 90% of Rainforest Carbon Offsets by Biggest Certifier Are Worthless, Analysis Shows." *The Guardian*, January 18, 2023, sec. Environment. https://www.theguardian.com/environment/2023/jan/18/revealed-forest-carbon-offsets-biggest-provider-worthless-verra-aoe.)

effectively stopping the country from making legally binding commitments to reduce its emissions under the pact. In the subsequent administration, President George W. Bush formally withdrew the United States from the pact. Alvaro Umaña said the protocol was "born dead" because just months before leaders negotiated the terms, the Senate had passed a resolution that barred the United States from entering into any agreement unless developing countries had similar reduction mandates.

Long-range targets set voluntarily by companies or countries became the common approach that is still in use today. An example would be a company choosing to cut more that 90 percent of its emissions by 2050 in striving for net zero. Yet Alvaro told me that long-range targets with large cuts don't work. "If you tell me cut 50 percent in ten years or 5 percent next year, I'll absolutely take 5 percent next year because that gets you going. The way the world figured out of this dilemma was to postpone it, to kick it down the road."

Sheryl expressed concern about fake projects and offsets that haven't fully delivered. There are projects focused too narrowly on carbon that have other consequences. There are investors who want to buy low and sell high, capitalizing on nature's services as they become more coveted. She was devastated that this great idea had become "perverted" in some ways.

I thought of many other examples I'd come across of that perversion. As some interviewees mentioned, there are equity and justice issues surrounding who profits from the sale of carbon offsets, in addition to a lot of potential for exploitation. In a mandatory system, the money that a company pays to purchase a credit goes to the regulator, a government, as compensation for the company's excess in emissions. I think of it as a form of permission, this mea culpa for not meeting a reduction that is required by policy. For example, the 2008 Climate Act committed the United Kingdom to reducing its greenhouse gas emissions by 80 percent by 2050 compared to 1990 levels. It was the first country to establish an enforceable carbon budget, which ultimately ensures emitters to reduce, remove, or avoid emissions in a variety of ways.

Revenues generated from offsets on the voluntary market, in contrast, tend to flow more horizontally, between companies and, in the best scenarios with forest carbon projects, to the landowners themselves. Landowners can lease their land to another party that develops a credited forest; the owner earns money from an annual lease, while the company might profit more later. Wealthy individuals can also purchase relatively inexpensive land and invest in a future forest, with hopes of doing something good for the planet while also recovering costs and profiting from the offset sales. Other arrangements include percentages a landowner might retain of the carbon sequestered, like owning some financial stock in the service the trees provide over time. There is no universal playbook to avoid perversion. The United Nations has new initiatives underway to address gender inequities in carbon trading; inequities in land ownership, particularly in the Global South, tend to reinforce gender inequities in land-use decision making and potential profit sharing. Then there's the overarching injustice created by offsetting with carbon removals instead of emissions reductions; people are dying in the highly polluted "sacrifice zones," while others continue emitting.[20]

"We also need so many approaches," Sheryl reminded me, snapping me back to the reality of multiprong solutions and progress. "If we kick-started this, then on balance, it's better than not having had it. Emissions could be even worse." She sighed. "I've just got to tell myself that." The situation reminds me of a chronic smoker who starts up an exercise program but doesn't stop smoking. Better than nothing? Sure. But does it fully alleviate the most threatening health risk? Not by a long shot.

Her honesty also softened me. "I'm trying to find my way in this too," I said. "That's part of my own motivation in this project. Where we are today and what people are doing now is not perfect by any means. The failed projects or fake ones are also dangerous if they're used to 'offset'; so is the idea that everyone in the world can do this. Because at some point, the supply can't meet the demand. If everybody wants to hit net zero, it can't all come from trees, and it can't all come from offsetting in general."

In a 2022 interview with a Bloomberg Green reporter, Mark Trexler, who developed the methodology for quantifying the offset with the AES CARE project, said, "If all the countries of the world have all these policies in place to reduce emissions and encourage carbon sequestration, then there is no room for an offset market." Policies would enforce reductions with the support of a committed, global community. Had there been effective regulation for carbon like there was for SO_X and NO_X, a credit system with a limited amount of credits for a strict budget could have already done its job, forcing emitters to innovate, and then faded away in some alternate history.[21]

But our Earth and all life sustained by its functioning ecosystems didn't get one firm budget with all countries implementing policies to regulate emissions. Instead, the current situation is largely an opt-in/opt-out/design-your-own approach. As Trexler told the reporter, the Paris Agreement, adopted in 2015, elevated the interest in offsets when countries began sorting out how they'd deliver on their own voluntary commitments. Each country could essentially take action against the climate crisis by announcing its own *nationally determined contribution* to the effort and then detailing how it could achieve its own plan for reduced, removed, or avoided emissions. The convergence of those commitments, along with net-zero-by-2050 goals that surfaced in 2019 and the Global Potential study that same year, triggered a lot more interest in planting trees.

Following the news about the AES support for the CARE project, the *New Yorker* cartoon had highlighted the mockery that persists—this sense that planting trees for carbon might absolve anyone of their pollution sins. *Time*, on the other hand, published a short article calling the initiative "a healthy environmental equation." The truth is probably somewhere in between.[22]

In March 2022, I received a cold email from Kristin Thayer, head of partnerships at Watershed, a company that helps corporations measure, report, and reduce emissions. It had just received $70 million in funding for its work on accelerating decarbonization. Kristin had found my academic profile and was looking for another perspective on an afforestation project with potential credits for a client. By the time we connected days later, Watershed had decided to pass. The low price point had engendered skepticism among its team, which was advising clients on pathways to net zero. There were doubts about *additionality*—the requirement that the credit purchase facilitate emissions reductions that wouldn't happen otherwise—along with other concerns, such as the likely duration of the carbon storage. Her inquiry made the boom, which I already knew was happening, that much more apparent to me; people like Kristin with a lot of experience in business development were scouring the planet for ecologists and foresters who could advise on how best to harness the potential of trees.[23]

Talking with Kristen felt like a crash course in the offset market. She spoke quickly and eloquently about the surge of interest in natural climate solutions and new players entering the scene. There were a lot of people without knowledge of how forest ecosystems function or of trees in general—how to plant them, grow them, or protect them. Companies wanted to buy up credits for future forests before they sold out. She identified other companies that were already known for snatching up low-price credits; she was skeptical of what those projects might deliver.

"We didn't start with a goal of focusing on high-quality markets for carbon," she told me, with respect to Watershed, "but that quickly became a clear need." She'd fielded inquiries from potential clients who were asking for the cheapest credits and noted that some strategies for purchasing carbon removals lacked a high bar for quality in project selection. "We try to work it out and help them understand that the more expensive budget is better." Kristin was targeting a range of $25 to $46 per tonne for reforestation projects.* If a client wanted high-quality credits, they'd come at a premium price: often more than $50 per tonne per credit for nature- or forest-based carbon removal. She talked about "supercharging reforestation" with respect to the innovators and companies, like Terraformation, that want to make growing forests and delivering their benefits more efficient and widespread. Kristin said that it was becoming increasingly popular to buy or prepurchase future credits from projects that were just getting underway. It was only the first quarter of 2022, and she was already having trouble finding high-quality reforestation credits that were still available.

I turned to One Tree Planted, one of the most recognized organizations in global reforestation, as I was curious to know how popular credits were in its project portfolio. In 2022, the charity supported over 300 projects that planted about fifty million trees with partners in seventy-four

* In the carbon market, these are commonly known more broadly as "ARR projects," encompassing afforestation, reforestation, revegetation.

different countries. I was surprised to learn that only a handful of those projects were trying to generate credits for the carbon that they'd sequester. I also asked Tom Crowther about the percentage of certified carbon projects on the Restor platform in 2023. He called it "a tiny fraction" of over 130,000 restoration projects that the global platform was tracking at the time. In July 2023, staff working at the world's leading greenhouse gas crediting program on the voluntary market, the Verified Carbon Standard Program, shared a list of afforestation, reforestation, and revegetation projects that had verified or retired credits. Verra, the nonprofit corporation that runs the program, had 97 projects underway and another 138 in development. Granted, what still seemed like a short list to me was from just one certification scheme on the voluntary market. There are others. Verra also had over 3,000 projects that included some component of growing new trees, but those focused primarily on other approaches, such as avoiding emissions through renewable energies.

A 2021 report by McKinsey & Company forecasts an increasing demand for carbon credits that could be matched in supply on the order of eight to twelve gigatonnes of CO_2 per year by 2030; the upper-end estimate of that range is equivalent to nearly a third of the total global carbon dioxide emissions in 2022. The report identified nature-based sequestration, including reforestation, as one of the major contributing categories. Although many efforts to increase forest cover around the world are still not aiming to generate credits, I took the information from One Tree Planted, Verra, and McKinsey as an indication that doing so may become more common practice in the future.[24]

Yet, around the same time that the high-quality forest carbon credit supply became relatively limited, media outlets like the *Guardian* and Bloomberg ran a series of articles that raised alarm bells about offsets in general and those coming from forests too. "'Worthless': Chevron's Carbon Offsets Are Mostly Junk and Some May Harm, Research Says," headlined one in the *Guardian*. Large-scale plantations and hydro dams were among the projects in Chevron's portfolio that had low environmental integrity; it appeared as if they weren't leading to additional reductions that wouldn't have happened otherwise or were even at risk of releasing emissions. In June 2022, Bloomberg reported that the vast majority of offsets on the voluntary market were coming from "avoided emissions"—projects that protect standing forests or provide energy alternatives to fossil fuels; data from a review, cited in the article, indicated that less than 5 percent were actually *removing* emissions. The reporters called offsets "a paltry offering."[25]

"The critical articles about offsets can be scary and create some doubt, but I think they've also been helpful," Kristin said when I asked her months after our initial call for a more official interview. From her perspective, they'd exposed many flaws in this evolving system of carbon trading as well as helped educate customers about why it's critical to vet projects and pay more for high-quality credits. "They make our customers come to us and say, 'I don't want to be buying from those types of projects. Help me find good ones.'" Some of her clients were paying upward of $1,000 per tonne for direct carbon capture that could store carbon permanently. However, forest projects still had the attraction of providing "permanent enough" carbon storage—meaning for over thirty years, maybe longer—at a more economical cost. The most important thing, she said, is helping develop high-quality projects, but "there's no one source of truth" for what a good forest carbon project is.

Part of what defines this current tree-planting movement—what I now think of as something new under the sun—is people endeavoring to increase forest cover with such a wide range of perceptions about what a forest is, could be, or should be.

Reading and hearing the critiques of planting trees for carbon brought me back to Bethanie Walder at the Society for Ecological Restoration and five-star restoration efforts. She'd called restoration an imperative in conjunction with conservation to move the world from net degradation to—at a minimum—net neutral. "But the planet is still unraveling at neutral. So, we must reach net ecological improvement."

Even if everyone snapped their fingers and magically switched off all emissions, there's still a lot more repair work to do. *Insetting*—the idea that instead of *offsetting* emissions, more change could come from within—might offer another pathway. Such insetting supports emissions reductions within a company's own supply chain, as opposed to funding external projects that might make reductions or removals elsewhere. Some companies, such as Nespresso, are doing this already with larger-scale versions of Rivah and Johann's ecology of place at their home in Hawai'i; coffee plantations flourish under shade and with stewardship of the trees and soil that sustain production.[26]

For me, the question had become much less about how to harness the potential of trees for carbon and more about how to galvanize people to do this work effectively, more equitably, and for a greater good beyond the sequestration service.

The willingness to pay for the hope that comes with the carbon trees sequester has reached unprecedented levels, but that willingness to pay is still not enough. We can't fail at the renovations. Without forests, new and old, the house will collapse.

13

GUARDIANS OF POTENTIAL

The heavy steel doors creaked as Jill Wagner, a consultant who specializes in seed banking, opened the end of the shipping container. I stepped from one world into another. Outside, the pink plumeria flowers, often used to make Hawaiian leis, were blooming. It was hot and sunny, and there were nursery tables with trays of grasses and other plants. The interior of the retrofitted container felt cool and refreshing, and it looked like a laboratory in meticulous order. There were seed samples everywhere. Labels detailed the species, date, and lot number or owner of each collection.

"Seed is always the property of the landowner," she told me. "I'm like a safety deposit box. I provide the space to bank and store the seeds properly."

We were standing inside the operating protype of an off-grid seed bank, which, in the original vision, could be shipped to the most remote reaches of the planet to facilitate proper seed collection, drying, and storage. Replicates had made it to sites in Kenya, Uganda, and Tanzania to build capacity for restoration with native species. Yet the cost of retrofitting containers and shipping them internationally was proving too high to create a global network of these local hubs.[1]

I marveled at the curious seeds of the *kamani* tree, a species native to Hawai'i that tolerates ocean spray and salty soils. The dimpled, spherical seeds resembled brown golf balls in an aluminum cake pan. Jill had baking sheets full of *'ahu'awa* seed, light tan and long grain, like basmati rice.

I scanned the names of other seeds: *koa*—the largest native tree species of the Hawaiian Islands; *mamaki*—a nettle; *ōhelo*—a native berry; and moringa. She'd carefully cleaned them with various tools, and they were now drying, a critical step to keeping seed viable for future use. When they are then stored at cold temperatures, many seeds can be banked for years, decades, even centuries.

"We have an unprecedented event in human history, and that is the UN Decade on Ecosystem Restoration," Jill said with a sense of pride and hope. "We have never undertaken restoring—*globally*—the Earth."

She was concerned about a lot of tree-planting efforts that involved using only a handful of species and others focused on planting fast-growing species. Those approaches had set things off to a rocky start: "Is that creating resilient habitats and systems for the next generation and the future? No, it isn't!"

In the many interviews I conducted, China's Grain for Green Program, which launched in 1999, was often cited as a classic example of the risks and trade-offs of monocultures. As the world's largest reforestation scheme at the time, the program aimed to reduce flooding and soil erosion; China became remarkably successful at reversing its forest loss to gain. A 2019 study using satellite data identified China as a leader in regreening the planet, and the increased leaf area was coming more from trees than from agricultural crops. However, despite the contribution toward reforestation, the program failed to promote biodiversity.[2]

Bethanie Walder at the Society for Ecological Restoration called those forests "biological deserts." She'd said Grain for Green demonstrated that people could "revegetate" at a great scale with support and under certain political conditions. "When you look at the number of initiatives that are happening now globally, we *can* plant at scale. There are issues with having enough plant materials and seed, but the question also becomes, 'Can we do this work at scale and in an ecologically and socially responsible manner?'"

I was visiting Jill because she'd spent decades collecting and storing seed and training others to do the same. She'd helped build a network of native seed banks across the Hawaiian Islands and saw seed supply as a major bottleneck to restoring degraded ecosystems on a global scale.

Yishan Wong had tracked her down in 2020; she was one of Terraformation's earliest hires and became chief of forestry as the staff expanded to nearly 100 employees. By the time we met, she'd left the company and was consulting again. She was working to launch a starter kit for people to build lower-cost seed banks and to create a school for online learning about seed banking to sustain plant diversity.

Given her expertise, she'd also recently received a request to select seed for the moon. The European Space Agency wanted recommendations for species to send on a mission to cultivate plants within a settlement or base. It was hard to refrain from laughing when I heard this. I could only think of the Hollywood movie *The Martian* with Matt Damon. (It's so good that I've seen it six or seven times, and I'm not even a movie person.) Damon's character, NASA botanist Mark Watney, gets left behind on Mars and fights his way to survival by figuring out how to grow potatoes. His best line: "In the face of overwhelming odds, I'm going to have to science the shit out of this." Now here was Jill, the real deal, dedicated to collecting a diversity of seeds for use on Earth and guarding them for our collective future; she was well poised to help "science the shit" out of a moon trial. The moon experiment was focusing on habitat before food, so Jill later picked a fern, *Nephrolepis exaltata*, which establishes well on lava and creates great leaf litter, and *ōhi'a*, the tree I came to know through my time with Keali'i Thoene, the community programs manager for Terraformation who had taken me to visit Kaupalaoa.

Seeming disappointed but determined, Jill told me that the business model for a broad deployment of seed bank shipping containers on Earth hadn't worked with Terraformation. "That doesn't mean that the problem doesn't still exist!" Jill wasn't abandoning the mission; she had moved on to work with another company, Viridios Capital, to improve the seed bank design. "We need to amass seeds for these restoration efforts," she told me. She was talking about the seed-supply problem that affects planting capacity and even natural regeneration.

Intermittent seed production or intense fire that burns the seedbed can slow the natural recovery of any ecosystem after a disturbance. Studies have documented higher mortality rates in some forests ecosystems, such as mangroves damaged by hurricanes, when they are restored via planting activities as opposed to natural regeneration. (Clearly, planting is not always best for a lot of reasons.) Natural regeneration can work if there are seeds stored in the soil, invasives aren't abundant, and people aren't pressured to use the land in other ways. Yet local seed may not always be present. Invasives can invade and dominate. Left alone in the warming world, many postfire landscapes will become grasslands. "For those places that would ideally be forests for social and ecological values, we have a choice to make," said Brian Kittler, who leads the Resilient Forests program at American Forests. "Do you resign it to a conversion and a loss of all the services that a forest could provide? Or do you try to get that landscape back on its feet?"

Climate-smart reforestation requires not only selecting species and populations that are more likely to survive the hotter and drier conditions but also thinning trees and using prescribed burns, for example. The hard work can reduce the likelihood that more intense fires will wreak havoc from forest floor to canopy and across landscapes. A former colleague of mine noted that such increasing climate stressors risk turning some of the carbon sinks into sources. The same threats also hold for seed. "Rather than discounting the potential of forests to sequester carbon," he told me, "these risks should be held up as another reason to work harder and invest more, not less."

Many considerations, including the shortage of seed supply, mean that people can't foster forests everywhere that there is potential in a decade-long endeavor. Don't get me wrong; I love the ambition, but I also appreciate a practical strategy. "We have to be selective about where and why," Kittler argued. "That's a values conversation as much as a science discussion. So, we have to look at the objective information from the science [to discern what approach is most likely to work where] and then also decide where to put our eggs."

Especially when we have limited eggs.

While Jill was working at Terraformation, she and her team had conducted an assessment of how many seed banks are needed and where in order to meet the global restoration potential (RP). The dataset published by the Crowther Lab for the Global Potential study had defined the RP. Jill's team skirted the critiques about tree planting in savannahs, grasslands, and shrublands by affirming the "right trees, right places" adage and noting that seed banking can support restoration of nearly any ecosystem type—wetlands, grasslands, shrublands, and deserts included.

The assessment wasn't only forest focused, but it did yield compelling results. Somewhere between two trillion and seventeen trillion seeds are necessary to meet the global RP; refining that estimate further is difficult. The number and distribution of seed banks offered a more accurate evaluation of on-the-ground needs for addressing the supply challenges. The research team documented about 400 native seed banks around the world. Jill told me that was also being quite generous, given "some seed banks are literally a refrigerator in an office." They also determined that tens of thousands of seed banks are needed to meet the RP in a sustainable way. That work requires using far more than the mere 100 or so tree species that people are typically planting around the world and having a steady supply of seed. Red spruce in eastern North America, for example, produce cones every four to six years; some species have gaps of a decade or more. Collecting during a mast year can keep a project going during the lull, but that strategy alone won't suffice, given the scale of ambition.[3]

Examining the seeds in her shipping container brought me back again to that freezer in Canada and the forty-nine million seedlings awaiting transport to their climate-informed destination. I'd been so focused on where those seedlings were going and why. Yet what it took to collect the seeds and grow them into their fragile but hopeful beginning was equally relevant. They'd come from a network of orchards that produce seed for planting common forestry species. Such a well-managed and intensive approach can yield massive seed quantities for a limited number of species, but it's not feasible for the more than 58,000 tree species that have been identified on Earth. Collection in an orchard might include using a ladder to harvest cones with a picker and walking tree to tree to collect the genetically improved and diverse seed. In an American forest with a few species present, traipsing around and climbing trees to collect native seed from many individuals can also be a relatively reasonable yet laborious undertaking. However, quantity is harder to achieve. Some operations even use helicopters to harvest cones. The collection job gets tougher in a tropical forest, where a team might scramble through dense vegetation all day to find only two trees of the same species. That tree species may be just one of hundreds of others present, and those two individuals may not even be producing seed at that time. Capturing the genetic diversity of one species in a primary forest requires sampling from a lot of dispersed individuals and collecting across their canopies. None of this work is easy, and it's often overlooked. Jill told me that people usually imagine the whole process simply starting in a nursery.[4]

Whether the goal is native species restoration, a monoculture plantation, or anything in between, upping the seed supply to meet ambition, even at a local project level, is a notable hurdle to overcome. A 2020 analysis led by Susan Cook-Patton (the self-described "glorified carbon account"/senior research scientist at The Nature Conservancy) identified 51.6 million hectares (about 128 million acres) of potential area for increased forest cover across the contiguous United States. That area is increasing as fires burn more forests, and the gap between what's available for reforestation and what people have the capacity and resources to reforest is also widening. Actively reforesting about half that potential with seedlings by 2040 would require 1.7 billion seedlings produced each year, a 2.3-fold increase over 2021 nursery production levels. A 2020 survey of nurseries in the contiguous United States found that roughly 20 percent of seedling are produced from wild-collected seed. Because collecting wild seed from as many individuals as possible is not a process of selecting for specific traits, it can help preserve biodiversity.[5]

From seed to seedling, sapling to tree, there are losses at every step of the way. More seedlings mean even more seeds from the beginning.

Prior to meeting Jill, I'd read other assessments of seed shortfall in comparison to tree commitments. An analysis of the seed systems in four Asian countries—the Philippines, Indonesia, Malaysia, and India—revealed "a mismatch between the growing demand for priority native species and the limited seed supply in terms of quantity and quality." The countries have committed to restoring 47.5 million hectares by 2030, and that will take more than 150 billion seeds. Similar supply issues for native species have been documented in East Africa, where tree seed centers have traditionally focused on commercial species, many of which are exotic. The mobile seed banks made by Terraformation, like the retrofitted container on Jill's property, had been shipped to sites that were starting to be ecologically restored with hundreds of native plant species in Africa.[6]

Seeds stored in a central location like Kew Garden's Millennium Seed Bank in Wakehurst, England, which stores the world's largest collection of wild plant seed, are an effective method of *ex situ* conservation. *In situ* conservation, in contrast, relies upon preservation of large and small tracts of forests. Botanical gardens like Kew in London, the Royal Botanical Gardens of Edinburgh (RBGE), or Brackenhurst in Kenya, which was using a container like the one I visited, also serve *ex situ* conservation. They're home to a great diversity of living "specimens" that may also produce seed. As the curator of living collections at RBGE told me, "Our mission is to explore, conserve, and explain the world of plants—and hopefully for a better a future." Better methods for seed storage in situ can also expand supply for local use.

The more I looked inside the black box of the global reforestation movement, the more inspired I became, which surprised me, given the complexity of everything I kept uncovering. There is no shortage of problems—failed plantings, conflicts over land use, shortfalls in capacity and resources, and faulty forest credits that don't deliver the promised carbon sequestration service, to name a few. Yet there is also an abundance of people willing and wanting to problem-solve the issues that arise and to keep working toward a more forested future. Inside the shipping container, I felt in awe of the care involved and the dedication to one dimension of this hope-filled journey. I was seeing the same across every dimension that I investigated.

I also sensed the rift between tree time and the results some people want trees to deliver immediately and forever. This reflects a fundamental tension between the desires of investors or stockholders and how ecosystems actually work.

"Restoration used to be very small scale because people were constrained by funding and staff," Jill Wagner told me inside her off-grid seed bank. "Now the funding is flying in, and the main thing on everyone's mind is speed." It's a new age for biology. "If you really look at a project in a holistic manner, you've got to ask yourself, what is your species list? What was the previous land-use model on this property? What was the ecosystem type before it was altered? And what are we trying to achieve instead of just going for speed? There's a lot of virtue signaling going on, but we have to do this properly."

When people don't have seed readily available, they might buy it elsewhere, try collecting it year to year, or get what they can from wherever to move a project forward. Those are very different strategies than one that requires up-front planning to determine what was previously present as well as what might survive into the future and then to create a seed supply accordingly. Climate change is already impacting masting, making the timing of seed production more variable or altering quantities produced. "If the mother trees are dying from megafires, then we are also losing the genetics and seed from those founders," Jill told me. "If we think about how on average a tree starts producing seed after seven to ten years, and we lose our mother trees, how are we going to collect?" Land-use changes, habitat fragmentation, climate, and extreme weather conditions also affect seed dispersal, and those stressors will continue to affect which plants may or may not regenerate on their own.[7]

"If you look at human history and you go back to the Paleolithic, we were hunter-gatherers, and we lived in small groups, and we moved a lot," Jill said. The shift to the Neolithic around 9000 BCE marked the agricultural era. "We started storing grains, creating permanent settlements, and populations grew in places that weren't conducive to growing food. I keep all this in mind because storing grain for food changed the course of human history." Yet only within the last century have people begun global seed banking for food security. She reiterated that never in human history have people taken a global approach to planting new forests and ecosystems.

Perhaps what people do in the years to come under the UN Decade on Ecosystem Restoration, and what lies beyond will change the course of human history again. There is that potential, if the result is an even greater shift in how people value standing forests—beyond carbon.

So, who is doing forest security? I thought. *Is anyone even calling it that?*

Meeting Jill and seeing the hundreds of thousands of seeds she had stored sent me on an unexpected search for what I could learn about the work of Russian botanist Nikolai Vavilov. She'd mentioned him briefly as the very important person who'd started the first global seed bank in Leningrad (modern-day Saint Petersburg) in the 1920s—not that long ago. I hadn't heard of Vavilov. I also hadn't adequately considered the parallels between banking agricultural seed to avoid famine and doing something similar for trees and other wild seed amidst these colliding climate and biodiversity crises.

As a young boy, Vavilov had witnessed the devastating effects of drought, insects, disease, and cold temperatures on crops. Nikolai (or Nikolay in some accounts) was only four years old when famine spread throughout villages in central Russia, threatening the health and survival of millions of people and killing 400,000. Ethnobotanist Gary Paul Nabhan, who retraced Vavilov's quest to end famine, writes, "Living through the famine of 1891 and 1892 had a profound effect on Vavilov. . . . It was not simply a historical fact that Nikolay learned about as he was growing up; it tangibly determined what was or was not available on the kitchen table of his home during his boyhood in Moscow." Reading about Vavilov reminded me that what we learn and experience, even as children, can motivate a drive for change; the young climate activist Greta Thunberg epitomizes this today. I can only protect my young son from knowing about the differences between his ecological baseline and my own for so long. At some point, the sheltering will feel like hiding the past and condoning the actions that still perpetuate environmental degradation. Vavilov chose plant pathology for his career because he believed it could benefit people in his country and around the world.[8]

Vavilov observed nature and noticed that wild crops have more resiliency. Collecting crop seed across five continents in more than sixty countries over twenty years, he and his colleagues built what was then the largest collection of seeds from cultivated plants in the world. Seed saving, or seed keeping, for future use stems from a long tradition of Indigenous people storing seed to plant for subsequent harvest, but building a global collection that captured such a diversity of crop species was new.[9]

"In my mind," Jill had said, "he was the first person who had that 'aha moment' about biodiversity."

His story has a tragic ending. Vavilov was imprisoned as those involved in genetics became one of the many groups targeted during the Stalinist purges. He died of starvation and left behind more than the tangible legacy of the seed collections. Nabhan writes that scientists today take as given that agricultural biodiversity "is the cornerstone for building greater food security for humankind." Vavilov was among the first to articulate this idea, and it ultimately cost him his life.

Without agricultural diversity, pests, droughts, floods, changing climate conditions, and more would cripple the food system. That same thinking applies to the future of our world's forests, but people are behind on the doing.

Now, about 100 years later, there are over 1,700 seed banks that create a global network for food security. Informal community seed banks have also become widespread; farmers manage seeds of mainly local varieties for local conservation and use. These community banks have been particularly effective in low-income countries, where access to quality agricultural seed may be limited.[10]

Yet a common misunderstanding about seed banks is that they are only for storage in preparation for crisis. Except for the Svalbard Global Seed Vault, dubbed the "doomsday vault" or the "Noah's Ark of seeds," all the others are supplying seed for use. Unmanned, located about 800 miles from the North Pole, and built into the side of a mountain, the Svalbard Vault safeguards duplicates of more than 1.2 million seed varieties. It has space for millions more. The vault stores the most diverse collection of food crop seeds in the world—on hold at subzero temperatures for hundreds of years, or until an apocalypse.

Built in 1953, decades before the Svalbard Vault opened in 2008, the US Department of Agriculture's National Laboratory for Genetic Resources Preservation maintains another impressive collection. I went to visit it in Fort Collins, Colorado; I wanted a window into the well-developed system for food crops. Designed to withstand natural disasters like hurricanes or flooding, the building is surprisingly nondescript. It is beige with thick cement walls and limited signage. Inside, countless steel doors with various locks and security systems guard the bounty. On my visit, bar-coded bags filled the shelves in the cold storage rooms; it was −18°C (about 0°F). In another room, there were tanks filled with liquid nitrogen for submerging other samples, as some seeds and other plant matter keep better at much lower temperatures. Scientists at the lab follow a systematic approach to testing seed viability. This ensures that the on-site collections can be used.

Touring the vaults with a plant pathologist at the lab got me thinking about the high-forest, low-deforestation (HFLD) countries that Matt Hansen had described as retaining our "big blocks—our greatest hopes." We had been talking about the relatively large, intact tropical and subtropical forests remaining and the fact that what the world might gain in terms of new forest cover could not equate to what it would lose with persistent deforestation. In addition to the collections that people create, the seed produced by our standing trees—whatever and wherever they are—will dictate the potential for any version of a forested future. I saw those big blocks and scattered fragments as a network of naturally occurring wild seed banks.

The fate of the "forest system" that any tree planting is creating depends on biodiversity too.

A global effort to grow more forests requires scaling resources across what people are now calling the *reforestation pipeline*—every step, from seed collection and storage, to nursery production and planting, to longer-term care and monitoring.[11]

In my eyes, if there is a present-day Vavilov, it might be Paul Smith, secretary-general of Botanic Gardens International, the largest plant conservation network in the world. He's also the former head of Kew Gardens' Millennium Seed Bank (MSB), which launched in 1996. ("This is all so new!" Jill had said of the overdue yet growing interest in collecting, storing, and using native seed.) Paul is trained as a plant ecologist; he grew up in Zambia, Botswana, and Swaziland, as his parents taught in various Central African countries. He first came to Kew in his twenties, when he was doing botanical inventories in Zambia and needed to identify species. Later, he helped set up collection programs with Kew in Madagascar, given his extensive field experience and knowledge of regional plant species. That work primed him to lead MSB, as the initiative began building partnerships and sending teams to make wild seed collections around the world.

Paul thought the term *forest security*, which I proposed, couldn't capture the scope of MSB's work. One problem he raised has to do with what a forest is, of course, and the fact that most definitions don't differentiate between "a man-made planted monoculture and something that's more akin to nature." MSB has also focused a lot on other ecosystems, such as savannas and deserts. But if we're taking forest to mean "primary" or "wild" forest or "naturally occurring with relatively minimal human intervention," then I'd still say MSB is at the forefront of forest security and plant conservation more broadly. ("We're going to start using that term now!" one MSB researcher told me, when I inquired again on site.)

The collection has about 18,000 tree species, and scientists at MSB lead cutting-edge research on wild plants and biodiversity. I was impressed by the X-ray machine and rigorous protocols used for assessing insect infestation and seed viability. One lab contains small rooms set to specific temperature and moisture combinations to mimic the conditions of seeds germinating in their natural habitats. The Society for Ecological Restoration makes Kew's species information available online for about 58,000 taxa (not only trees) studied to date. With species-specific details on storage and germination, for example, the database is just one of the many resources available to facilitate restoration.[12]

"One of the problems we're seeing with tree planting," Paul told me, "is there's nothing in it for local communities, and we're seeing these imposed models. What happens is a company will come in and put clones in, and then they'll aim to leave them for thirty years while the carbon accumulates." If people plant native species, they need local provenances of those species and local expertise on how to grow them too. However, even the more restorative projects using native species will commonly bring in experts from elsewhere. "We have to change that paradigm so that it's local biodiversity experts who are making decisions and it's local people who are more involved in value chains," Paul said.

Despite the extent of the MSB collections, the current use is more for research that supports restoration than for restoration itself. As one scientist there told me, "Ten thousand seeds per species sounds like a lot, but it's a drop in the ocean for a large-scale restoration program."

A global network of local seed collectors, banks, and nurseries can also create hubs of local knowledge, empowering more people to benefit from and direct what any forest might become. At least right now, that vision has yet to be fully realized in the race for a forested future. But it too has potential.

14

CREDIBLE LINES OF ATTACK

Scotland is quite small and nowhere on the list of top-ranking nations for tree-restoration potential, but it remained on my radar because of its historic reversal in forest cover that had sparked Alexander Mather's early work on such transitions. Sometime in 2020, a colleague sent me an article from the *Guardian* about the United Kingdom's need to embrace nonnative species to sequester more carbon. The lead read, "Non-native conifer plantations have long been the scourge of conservationists—blamed for wiping out woodland species and disfiguring landscapes. But exotic conifers will be better at tackling the climate emergency than much-cherished broadleaved woodlands, according to the outgoing chairman of the Forestry Commission."* Sir Harry Studholme, the outgoing chairman, described the planting of Sitka spruce on peat bogs in the Scottish Highlands in the 1970s and 1980s as "a tragedy" for other reasons than the disdain for the nonnative plantations. The land-use change that occurred back then had shattered the integrity of peatland ecosystems, which are powerful carbon sinks themselves, in exchange for timber production. The plantations degraded habitat for migratory birds and other species too.[1]

* Since its establishment in 1919, the Forestry Commission has been responsible for managing and expanding publicly owned forests in England, Wales, and Scotland. Such authority of forests devolved relatively recently. In 2013 and 2019, separate entities were established in Wales and Scotland to carry out similar duties.

Decades later, as the *Guardian* story revealed, there were new arguments that Studholme could support for fast-growing conifers, such as Sitka spruce.* Subsequent articles drew attention to a so-called green rush with investors buying up rural lands. Alongside corporate "green lairds," the *Atlantic* reported, "wealthy individuals are buying large properties in hopes of restoring wilderness and regenerating carbon sinks." I recognized this as a very different kind of prospecting, but it has some historical parallels.[2]

The news in the United Kingdom reminded me of terms I'd heard in my interviews that characterize concerns about forest carbon offsets more broadly:

The Wild West, in reference to the surge of new players entering the scene with no consistent framework for quality projects. It means anything goes. One person involved in sourcing offsets for companies told me, "There's no good calibration system. A credit from one project is not equivalent to a credit from another project." She said that means it's difficult to choose what kinds of investments to make. A tonne of carbon here might be a tonne of carbon there, but the actions that achieve one tonne of sequestered carbon and other outcomes—for biodiversity or local livelihoods—can be dramatically different between projects. Determining the quality of a project requires assessing more than the amount of carbon it might store.

Carbon cowboy, someone chasing after cheap credits in the race to the bottom, which probably could have been avoided by establishing a better carbon pricing system from the beginning.

Crystal ball gazing, meaning there's a lot of speculating about how projects will pan out and how the offset market will evolve.

---

* Woodland in the United Kingdom is technically defined by a minimum area with stands of trees providing, or with potential to achieve, at least 20 percent cover. However, in casual conversation, people sometimes refer to mixed broadleaf forests and more ancient forests as *woodlands* or *native woodlands*, whereas *forestry* infers conifer plantations. Always distinguishing the two is a convenient way to get around the whole what-is-a-forest conundrum.

As Steve Scott, area director at the Forestry Commission in England, told me, "The carbon that the tree puts on in terms of girth and height is real, but the trading of that and how that becomes an asset feels no more real than the dollar in my pocket. It's a promise to pay. A tradeable asset." How people value that asset will shift too.

Carbon colonialism, typically defined as the ability of wealthier countries to outsource emissions to less wealthy ones. However, I'd heard this one used more commonly to flag the general practice of outsiders coming in, taking over land for the purposes of carbon sequestration, and running over the interests of local communities in the process.[3]

The colonialism term had surfaced with respect to companies from high-income countries financing forestry projects in the Global South and severing local relationships with the land. But I was learning that any change in land use that is motivated by new interests or values in a global economy could also run this risk.

In the United Kingdom, many areas with potential to be forested had trees at some point but not in recent years or even within the last century. So, the efforts to increase forest cover count as new woodland creation or afforestation, establishing a forest on land that was not recently forest, and that tends to be more contentious than reforestation. Creating a forest where one hasn't stood in living memory competes with perceptions of what the land and its uses should be. Even if I close my eyes and imagine a few quintessential views of Scotland or England, I see rolling hills and open land, gnarled oak trees and grazing sheep. I can understand how altering the current norm can incite resistance. Will planting trees push sheep farmers off the land?

"Culture is really important," Sir Harry said when we connected through a video call. "I don't think one should underestimate that." Culture is deeply tied to how people use the land and perceive their relationship to it across generations. Another project developer explained that people in the countryside "want to see fields not trees, but everyone wants to have a global picture of more trees." There's a tension.

The media coverage surrounding the green rush, this race to buy up land for its prospective value as forest, failed to represent the breadth of approaches underway for increasing forest cover in the United Kingdom. While Scotland has relatively larger tracts of land that may be suitable for forests, the vast majority of afforestation projects pursuing credits throughout the United Kingdom occur in small areas—ten hectares, on average. The coverage also seemed to sensationalize a race to acquire land where forests can grow and for planting species that deliver quickly on carbon too, as if every project is playing out that way.[4]

I was intrigued by the Woodland Carbon Code (WCC)—the standard for any woodland creation (including plantations) that generates carbon credits in the United Kingdom. One Woodland Carbon Unit is equivalent to one tonne of CO_2 that has been sequestered in a woodland and verified by another party. Since the WCC launched in 2011, before any country officially targeted net zero, over 500 projects have been validated for the carbon they sequester. In Scotland, that number is split between forests managed for timber and "minimal intervention" woodlands that are primarily managed for biodiversity or aesthetics, though that balance could shift.* As of June 2023, another 1,400 projects were in development. Only companies that emit carbon within the United Kingdom can use WCC units to offset their UK-based emissions, so the scheme is set up to drive positive change in the homeland instead of outsourcing the fixes elsewhere. As one Brit involved in forest carbon initiatives overseas told me openly, "The British do things like the Carbon Code properly; they don't always get it right, but at least you feel like it's being organized by adults!"

I also saw the United Kingdom as a microcosm of the global challenge to manage the suite of human pressures on limited land. Steve Scott at the Forestry Commission told me, "Whether it's for food security, biodiversity, residential development, business development, or for trees—and there's more in there too—we are all effectively competing for the same bits of land in one of the most densely populated places in the world." In recent years, the United Kingdom has been the second-largest importer

---

* The distinction is not quite so binary in practice because even the projects focused on managing forests for timber always include some area that allows natural processes to occur and supports other benefits, such as biodiversity.

of timber in the world, behind China. Caroline Ayre, a forester of nearly thirty years and former England national manager at the Confederation of Forest Industries, said, "We have a gold-plated international standard by which we are compliant within the UK for forestry and forestry management. So, if we continue to import timber from countries that may or may not have the same high standards because some people don't like us planting spiky trees on farms, we put our own sustainability standards at risk. We need to plant our own." She says the United Kingdom has a responsibility to produce its own timber, and that means planting conifers.

Interest groups and citizens who want to support more biodiversity and also perceive woodlots of broadleaf species as more characteristic of the landscape oppose the plantations. When we spoke, Sir Harry said the idea that "coniferization" is a terrible crime is also deeply engrained in the public consciousness, despite the need for timber. I had never heard that term for full conifer takeover, but I loved it. Distress about favoring conifers at a global scale is also widespread.

The UK government is targeting a rise in forest cover from 13 percent to 19 percent by 2050. A report from the government's Climate Change Committee estimates such an increase in forest cover requires 0.9 million to 1.5 million hectares. The amount of carbon that those new forests could sequester annually is equivalent to about 4 to 7 percent of the United Kingdom's total carbon dioxide emissions in 2022. Some of that land would need to come from changes in grazing intensity and consumption of beef, lamb, and dairy. (Again, so much for the simplicity of, "Let's plant trees.")[5]

Not everyone aiming to increase forest cover will pursue carbon certification in the United Kingdom either. Foresters, conservationists, and citizens have been divided on the best planting strategy to meet the cover targets. John Weir, a retired forester from the Forestry Commission, said, "We should be looking for a balance between very fast-growing species—which are not native because we don't have fast-growing native species—and native forests." Some advocate for creating more native woodlands that can offer many benefits over time, eventually becoming the ancient forests of the future. Restored mountain woodlands in Scotland, for example, will sequester carbon very slowly, but they'll protect farmland and infrastructure from landslides and other effects of extreme weather. Others want nonnative plantations that can be harvested for timber and sequester carbon more quickly.[6]

I was itching to understand the perspectives of the project leaders operating on both ends of that spectrum.

I went searching for a project that exemplified a relatively extreme interest in fast-growing species, demonstrated an innovative strategy for changing land use from farm to forest, and created plantations. I came across a *Paulownia* planting endeavor.

"Every generation or so, there's a new hope that springs forward in forestry, and you can see patterns of that on the landscape," Steve Scott at the Forestry Commission told me. He was talking about the British landscape, but I thought of Virginia Norwood and the satellites, drones, and other emerging technologies that will also document the patterns of this larger movement around the world in the years and decades to come. "In the 1950s, everybody thought poplars were going to be the new thing, and there's that legacy in the landscape," he explained. "In the 1970s, there was interest in planting Norway maple at a large scale, and you can still see that in that landscape. Roll forward to now, and *Paulownia* is the latest in that kind of new hope. It's too early to know whether this is going to be successful or not."

Paulownia, a genus of hardwood species that is native to China, was originally named *Pavlovnia* in honor of Anna Pavlovna, queen of the Netherlands. She loved plants that came from other parts of the world. With their enormous heart-shaped leaves and trumpet-shaped flowers, these trees were a favorite. Also known as the princess tree or the empress tree, it's one of the fastest growing in the world. Almost a real-life Jack and the Beanstalk, *Paulownia* reaches maturity in seven to ten years. It stands proudly on the landscape at heights that other species would need decades to reach. I found photos online of the Phoenix One, a cultivated hybrid between *P. fortunei* and *P. elongata* that documented eleven meters of growth in a little more than two years.[7]

Published literature describes the cultivation and utilization of its timber in Asia as far back as 221 BCE, but the genus has garnered more attention recently for its prowess in carbon sequestration. Some estimates show growing an acre of *Paulownia* as equivalent to taking about eighty cars off the road each year—approximately 200 cars for a hectare. Other studies have deemed its use in agroforestry as ideal; it serves as a short-rotational tree crop that helps absorb emissions.[8]

"Don't get me wrong, I love a good oak," Nigel Couch, the managing director at the UK company Carbon Plantations Limited, told me. "But we believe in six or seven years, this tree does what it takes an oak twenty-five years to do. I'm not suggesting it replaces native trees, but it can sit alongside what exists today."

Some people I interviewed called the *Paulownia* a very "alien-looking tree" in the British landscape. Every fall, it creates a thick layer of organic matter on the ground when it drops its leaves. Nigel said that one November, someone living near their fields phoned the farm manager to inquire what had happened to the trees. "Has someone stolen them?" they enquired, startled. The seasonal change is so sudden and dramatic, leaving bare poles on the landscape until the spring bloom comes again.

The Phoenix One, the hybrid that Carbon Plantations uses, is sterile. Unlike Pavlovna's trees or the invasive *P. tomentosa*, which spreads extensively in the southeastern United States, the hybrid won't flower to produce seed. In theory, it stays where people plant it.

It can be coppiced, or cut back to stimulate growth, eight to ten times over eighty years: "You cut it down to its knee, literally about a foot above the ground," Nigel explained. "Then, you find the straight stem leading it, and you train it to go back up again." That also means the root ball underground remains undisturbed. "All the good work you're doing underground stays in place with that carbon sink." Other points of attraction in a warming world: it tolerates a wide temperature range, and although it requires watering during its first couple of years, that need wanes after establishment.

By the end of 2022, Carbon Plantations had planted 200 hectares of *Paulownia* in Suffolk, England, on lands used previously for farming root vegetables. I asked many foresters and people involved in carbon certification in the United Kingdom about their impressions of the Paulownia plantation. John Weir's response struck me: "People are often screaming about a monoculture, but a field of potatoes is a monoculture.

They're not even native; they're alien here too. So, the argument doesn't quite square up."

Carbon Plantations sourced over 1,000 hectares from about twenty different landowners who wanted to diversify their farms. Carbon Plantations didn't buy the land; the company leases it. "Effectively we are tenant farmers, but farmers of trees instead of wheat or root vegetables," Nigel told me. In what he calls "a symbiotic relationship" between the landowners and the company, the landowners receive an annual payment for the use of their land over decades, and the company retains the carbon credits along with the ability to sell them for profit.*

The leased land now comprises 75 percent *Paulownia* and 15 percent native species, leaving some open ground, and Carbon Plantations was already expanding to another 250 hectares in 2023. Nigel said that even if they hit 1,000 hectares, that would still be a "little dot on this part of the world." A big effort on that little dot may yield some new information about what's happening underground, because *Paulownia* trees are massive rooters.

"Within the first year, we've had some roots grow down to 1.8 meters," Nigel told me. He and his colleagues estimate that 70 percent of the carbon that the trees sequester will be in the soil; calculations that have focused on what happens above ground may be underestimating *Paulownia*'s full potential. As part of the project, they'll track below-ground carbon and nutrients using the latest technologies that combine digital scans, captured by sensors on a light-weight all-terrain vehicle, with soil-sample data.[9]

John Purslow, an agronomist working on the project, sent me preliminary data showing changes in fungi and bacteria in the soil within the first year. They were building up the understory, using arguments like those I'd heard from Rivah and Johann in Hawai'i for a very different regenerative

* In the UK system, there are woodland carbon units (WCUs), which have been validated for the carbon they sequester, and pending issuance units (PIUs), which are more like promises to deliver. PIUs are issued after planting, and they should convert to WCUs over time if everything goes as planned. Earlier "vintages" are more valuable than later ones, because, assuming the PIUs deliver, whoever buys them can use them to claim CO_2 reductions sooner. A markets advisor told me that a PIU with a vintage of 2025 could sell for 100 times more than one with a 2100 vintage. A super-fast-growing species like *Paulownia* is more attractive than a slower-growing conifer in this respect, because the units will convert from pending to validated ones quicker. *Paulownia* makes for a relatively early vintage.

effort. Restoring life below ground can stimulate the life above it too. Purslow told me that he has a "great belief in the symbiotic relationship between fungi, nutrients, and plants" and described insects as the "glue of ecosystems."

Nigel called the effort a "commercial operation" and one that, in its purest sense, "is a capitalist way of capturing carbon." But even so, they aim to build back natural relationships that have faded in a new kind of functioning ecosystem. "Now if you say to me, 'What's the biggest issue about climate change right now?' it is the fact that we haven't got enough time." He said people need to think outside the box to accelerate solutions. "We've got to consider ways to facilitate faster carbon capture in the natural world, and I think this tree needs to be given the opportunity to do that."

The ecologist in me was and still is hesitant, even resistant. Like a countryside Brit cringing at coniferization, I find the idea of trying to promote natural relationships between fungi, bacteria, insects, and birds through a carefully bred tree that is planted in lines counter to what I know and trust. But I also appreciate another take on "right trees, right places" and a sound experiment. I'll take plantation trees on old farmland before suggesting that they replace primary forest. Carbon gains will be faster and quicker with *Paulownia*, but the risks for unintended consequences may also be higher than they would be with native species. Time tells all.

"I think this is a very credible alternative to existing lines of attack on carbon capture," Nigel said. "I'm not suggesting it takes over in any shape or form, but I think it sits well along with what we're already trying to achieve."

The word *mélange* came to mind. What people are rebuilding and creating in this movement may become a mélange of green across the Earth's surface, each patch distinct and shaped by what was possible in time and place. The *Paulownia* plantation is a bit of trial, but then again, isn't it all?

"People perceive forestry as, 'Conifers all the time,'" said James Hand, an operations forester for Forestry Land Scotland, the Scottish government agency responsible for managing national forests and land. "So, the media blows that up sometimes too. 'Oh yeah, forestry coming in—big Sitka spruce everywhere.'" He spoke quickly. I had to decipher words through his accent. "These big native woodland creation projects, like Loch Katrine, just doesn't get the same coverage, but there are also carbon projects that are really benefitting the environment, habitat, and connectivity."

During a visit with my husband's family in England, I'd taken the train from London to visit Loch Katrine, the site of one of the earliest carbon-certified forest restoration projects in Scotland. The air felt soupy. It was pouring rain. James had picked me up in Stirling, a city in central Scotland with a medieval castle that stands on craggy volcanic rock. He fell into tour mode immediately, narrating our drive across layers that had formed during the Carboniferous. As he spoke about the coal deposits, I thought about the history buried below and the hope planted above.

Scotland was covered in glaciers until the end of the last ice age. The retreating ice carved the valleys, or glens. Despite the more recent perceptions of "coniferization" that emerged in the twentieth century, conifers like junipers were among the first pioneer species. When people began farming some 6,000 years ago, pines were the first to go. There was a period when "normal" wasn't fields but forests of conifer and broadleaf trees.

I'd come to Loch Katrine to see a project that had originated from a few motivations, including monitoring and delivering carbon benefits over time. The lake, or *loch*, as the Scottish say, supplies freshwater for the central belt of Scotland. The water goes to Glasgow and onward from there. Loch Katrine and its surrounding land lie within the nation's first national park, which was formed in 2002. The banks were used to host the largest sheep holding in Scotland; some accounts indicate it was the third-largest sheep holding in all of Europe. Yet land use needed to change when an outbreak of a waterborne disease was traced back to parasites carried by the lambs. In the first ten years of a 200-year project, more than 2.5 million trees were planted with funding from British Petroleum (BP) to create woodland again. Those involved early on knew expanding the temperate forest and restoring wetland areas would help protect the water resource and deliver other benefits too. When the Woodland Carbon Code came into effect, the project entered the credit system, with BP holding the rights to the early units.[10]

"What we want is natural regeneration," James said, as we suited up in waders, wellies, and raincoats for a walk. "We don't want to plant everything all the time. The trees want to come back; they need a release." They were now planting seed islands—forest patches in strategic locations that could facilitate natural regeneration over time. Deer tend to eat small saplings along with other forage, so in what was seemingly one of the most controversial aspects of the effort (minus the whole idea of putting trees on the landscape in the first place), they'd been paying "stalkers" to kill deer as a means of cutting down on browsing damage. That wasn't so well received by neighboring landowners who were taking paying customers to hunt deer that were getting shot to protect saplings instead. But when the deer population is too high, new shoots get eaten by the hungry herbivores in the winter. The seedlings can't grow into trees.

We trudged up a hill, stepping over tussocks and slipping in mud. It was cold. Areas with planted Scots pine and rowan appeared like specs in the bucolic landscape.

"We don't want to go straight into what we want it to be," James said, as he described the project plan. They had been planting shrubs, willows, and birches knowing those species would improve the soil and create microclimates for others to establish later. "In thirty or forty years, we'll start seeing the woodland come, but I anticipate it will take sixty or seventy years until we're seeing a really established woodland." James raised his hand to about shoulder height. "If I come back here when I'm eighty, and the trees are about wee high, then we did well."

Something about the contrast between a 200-year, slow-growing native woodland and a fast-growing *Paulownia* plantation illuminated how much a project's priorities dictate from the beginning.

"This is all a change in land management and in what people are trying to get out of the landscape. It's a generational thing," James told me.

I thought of the five-star projects on the restorative continuum: the farther right a restoration initiative is on the continuum, the greater the biodiversity, human health, well-being, and climate benefits. Putting an economic value on carbon has enabled people to create more forests, but we can't lose sight of all the other values that can come from any effort.

"Would we be doing this work if the carbon money were not there?" James posited. "Probably not." Nevertheless, the heightened interest in carbon is a bit of a double-edged sword. Funding was accelerating planting and the development of new projects, but labor costs had gone up with the demand in Scotland. "There are landowners who have wanted to create more woodlands for a lot of reasons. Now there's some money. There's an incentive to do it. People are quicker to say, 'Let's jump on it.'"

Standing on that wet landscape of muted grays and green, I thought about the sheep that roamed, the farmlands of the past, the cover of conifers and broadleaf trees, and the ice. I also thought about the *Paulownia* project and the very reasonable assumption that far more is happening underground than people know. I asked James about the soil carbon at Loch Katrine.

James told me that the carbon calculations are based on the productivity of the trees, considering the species, what we know about their growth rates, and other factors. "The calculations are based on some things we understand well, but then there's a lot of guessing and room for error too," he admitted. "We're not even considering soil carbon, and that's something I've always wanted from the start."

I appreciated the honesty that the project might be overlooking soil carbon, and I didn't fault him or anyone else running similar calculations. Scientists always want to discover more, placing one building block of knowledge upon the other. However, repeatedly saying, "We need to study *x* or *y* before doing *z*," can also stall action. I'd talked with researchers and project developers in other countries who are also trying to understand the soil dynamics; Pedro Brancalion, for example, a scientist and expert in tropical forest restoration, has had a team of some forty people working across about 700 sites in Brazil to assess outcomes of different restoration approaches. They have been studying the carbon outcomes both aboveground and below and, as of March 2024, had collected data from 50,000 trees and 1,000 species.

"It gets weird when you compare baselines and think across different points in time," I said to James. "We are just one more blip! Some years from now more people might be focused on soil carbon. Then, what? Maybe there's an ecosystem or land use that we didn't even know stored so much carbon. Then it's 'Let's make more of that!'" All of it seemed like another argument for more holistic restoration, wherever and whenever possible. Best not to sacrifice one ecosystem value for another when there's still so much that people are learning about how they all relate.

We got back into the van and continued driving around the lake. I was stuck on the tension between carbon money enabling the work and the doubts surrounding the accuracy of the calculations. "Here's the hook," I said. "When you get to the act of offsetting, then it is absolutely critical that what people are doing is delivering the amount of sequestered carbon that they expect. That's where it gets tricky for me." James agreed. Attempting to compensate for the carbon emitted in one place with carbon removed at another is attractive. However, if the emissions do more damage than the sequestration efforts can fix, we've still got a problem.

In recent years, some teams within the Forestry Commission have tripled in size. Yet even with all the interest and effort, the United Kingdom hasn't been hitting its planting targets. In Scotland, 2019 was a peak year for forest creation; published rates show a subsequent decline. One Scottish consultant told me, "Two or three years ago, there were properties changing hands for large sums of money. You got the feeling that any estate could go into a credit scheme." But the fact that they aren't all going forward suggests to him that "there's something not quite right about the whole thing, or it'll just take more time."

Perhaps Steve put it best: "It's an evolution. It might become a revolution if we see whole land-use change, but I'm not seeing that yet." Even so, he says that farmers and landowners continue coming forward to explore the options for adopting different practices on their lands and pursuing economic incentives for nature-positive outcomes. They're expressing interest in woodland creation, carbon credits, and even new schemes for biodiversity credits that may become available.

Glasgow was foggy and damp. I watched my footing on the slippery sidewalks, treading lightly over mossy patches on the morning of my departure. Given gaps in train service, I'd opted for a flight back to Matt and Calder in England. As I walked to my gate, a blue British Airways sign grabbed my attention. In big letters, it read, "Did you know our flights within the UK are carbon neutral?"

I took a deep breath. *Really?* I thought. I was already feeling guilty for flying, and I recognized that I knew too much at that point for the sign to yield much relief. I don't want to convey the wrong message. I do applaud the mission. Yet the sign also triggered a series of rapid-fire questions in my head. *How much of that "neutral" comes from emissions reductions? How much is coming from offsets? What kinds of offsets?* I didn't know whether the company's plan included forests. I later confirmed that it did, with forest protection in Cambodia and Peru, for example. But at the time, I was thinking, *If the plan includes forests, is the carbon sequestered truly 100 percent fungible with the carbon emitted—and forever?*[11]

I chastised myself for the critical voice and played devil's advocate. *If I could snap my fingers and do away with the interest in trees for carbon sequestration, would I?* No. That is not the take-home message.

I think some good is coming from this, and more is yet to come. It won't be everything that everyone involved wants or expects. There will be some problems, as we have seen already. Time will tell those stories and what we are able to resolve too. Addressing the current climate crisis by halting emissions can't restore balance completely either. Even if someone waved a magic wand and stopped all fossil fuel emissions this late in the game, those emissions reductions still wouldn't be a silver bullet for stabilizing the climate system. We need to draw down more greenhouse gases that have already accumulated in the atmosphere to help solve the problem. Trees can help. A colleague of mine at Stanford says, "It's all buckshot at this point," referring to many small balls fired from a shotgun at once instead of a single silver bullet. I don't want a broader understanding of the many complexities surrounding tree planting to be another factor that stifles effort. I want it to help people do the hard work of growing trees *better* and keeping them—for more than the carbon that they sequester.

As my plane soared into the white abyss, I felt frustrated. With all the research I'd done, the people I'd interviewed, and the projects I'd visited, I was still struggling to see what "better" could be.

I'll get there.

15

LUCK FOREST

I gripped the handle above the passenger-side window with my right hand, held a recorder in my left, and braced myself for the rugged ride to the finca. It had rained heavily overnight on the Azuero Peninsula, raising some doubt as to whether the road was passable. Andrew Coates, who was driving the Land Cruiser behind us, was determined. Then again, Andrew had driven across the Sahara Desert and had twenty years of experience running projects in remote tropical places, where nothing quite works out how one might expect. "I quite like a high degree of chaos," he'd told me over breakfast. I'd come to Panamá in search of some version of growing trees "better" and keeping them over time.

Someone had described Andrew to me as the "real deal Indiana Jones" before we'd met. He fit the bill. He was dressed head to toe in khaki, with a faded handkerchief tied loosely around his neck. A snug, cloth sleeve on each calf extended over his pants and the tops of his leather boots, sealing off access by the bugs and critters we'd encounter. Malcolm Porteus González, project manager for Latin America and the Caribbean at One Tree Planted (OTP), and Ross Bernet, monitoring manager with the organization, accompanied us. When Malcolm and I had spoken about the recent boom in reforestation and efforts that OTP was supporting, I'd discovered that he too had plans to visit the same region in Panamá. OTP had given $800,000 to help kick-start the project.

That morning, I was riding in the lead truck with César Zambrano, a local Panamanian and the project's reforestation technical lead. Some people called him Céso or Césa, cutting off the *r* in more free-flowing Spanish than my own. César had successfully made the trip to the same farm the previous day, but he was now skeptical, given that the soils had turned to mud overnight. He was wearing a baseball cap bearing a bright red apple. His eyes lit up when he talked about his fruit trees. His English was better than my Spanish, but we spoke in a mix of both languages to find fluency together.

We drove slowly behind a white cow and waited for the animal to clear the road. I could see every rib arching across its chest. Hip bones protruded from its starving body. After a long and especially hard dry season, the wet "green season" had finally arrived. Two weeks prior, the now lush lands had been brown and cracked. Andrew said the contrast reminded him of when he had lived in East Africa; the buffalo would gather on the savannah to wait and watch for the grass to grow as the rains returned. Here, too, the grass could grow inches high in a matter of days.

Much of the tropical dry forests across the peninsula had been cleared decades before for cattle farming. Tropical dry forests are generally found between ten and twenty-five degrees latitude, north and south of the world's tropical rainforests. They are among the most threatened ecosystems in the world, with some studies showing that less than 1 percent still have a very low human influence. On the farms that surrounded us, livestock had been more profitable when the productive land sustained more cows. César told me that with the dry season intensifying and the lands degrading over time, most farmers could only keep one cow per hectare. There wasn't enough meat on any cattle I'd seen to make much money at all.[1]

I'd learned that farmers in the region could make $35 to $100 per hectare from cattle in a good year. In a bad year, like the one they'd just experienced, that income was unreliable. With reforestation, they'd earn $50 per hectare for thirty-five years, plus a fixed percentage of the gross income from credits sold on the market, as well as additional money up front for the site preparation, planting, and maintenance: $300 per hectare for years one to three and $200 for years four and five. The project developers were aiming to generate five million credits over the first thirty-five years; then the landowners could renew their contracts to keep growing the trees and earn a much higher percentage of the value of the additional carbon sequestered. Other experts I'd interviewed had shared the perspective that local people need to make a business out of their service to nature with other nontimber products so that they can survive after a big restoration project. "People are often left to figure it out once the funder is done, and they still need to live and have some income," Jill Wagner, the consultant who specializes in seed banking whom I visited in Hawai'i, had shared. Sometimes a funder is done in a planting season or a couple of years. Sharing the profits from credit sales is a way of incentivizing longer-term care and bringing more profit to the landowners over time, a far cry from a one-time property sale that pushes local people off the land. Those involved on the Azuero Peninsula also had ideas for developing other income streams from the forest over time; cacao and vanilla plants could grow with the native species they were planting.

I whistled out the window, and the cow climbed the bank beside us. César accelerated to get a strong start on the hill. Our tires spun, struggling for traction between puddles and rocks, but we reached the summit. He slammed on the brakes as we looked down a dramatic descent.

"Whoa!" I exclaimed, feeling the anticipation as if we were hovering over a roller-coaster drop. "Really?" I added, then regretted expressing any doubt.

César pulled my attention away from the road. "See this bare hill," he said, gesturing to the expanse before us. "This is what we're going to change!" I looked across the landscape, a verdant vista of grasses with patches of trees. "The idea is to transform all of this. I am pretty confident in the project. All of this has to be forest."

The windows were open, and we were dripping with sweat. César repositioned his foot on the pedal and then crossed himself, head to chest, shoulder to shoulder. I imagined he was praying for safe passage, but it also felt like a sign of devotion to the land, his community, and the forests of the future. My mind wandered for a moment back to the cross in the tree in Chile, that monument to the past, and then to those stepping-stones I had walked with Calder at Kew Gardens: What—do—plants—need—to—grow?

Surely plants were growing at the finca but not the *forest* that many people and other animals wanted and needed. Simply taking cattle off the land, fencing it to stop any grazing or other degradation, would allow for some trees to regenerate naturally from existing seed sources. The Scottish geographer Alexander Mather had documented that type of forest transition long ago; forests returned when people abandoned land in rural areas as they migrated to cities. "The forest will come back itself," Andrew had told me during our first phone conversation before my visit. "We just need enough key species to kick-start it. Then it will go off on its own." Endeavoring to replant the native species could accelerate such a reversal.

Forest clearing in the past had exacerbated the hot and dry local conditions, putting an additional strain on local water too. César said he was lucky now. "I tell people that we are making the forest a better place for you. I am living my dream."

Over the course of days, I would learn a lot more about how the project might achieve native restoration at scale. I'd visit recently planted lands and backyard nurseries. I'd talk to landowners who wanted to reforest and community members who collected seed. In an outdoor classroom, I'd listen to young children recite species names, and I'd watch them carefully transplant tiny saplings. I'd sit beside the farmer, Hector Frías, who owned the finca we were trying to reach, as he signed the first agreement for planting trees and the sale of carbon credits in the region. Carbon, I would realize, can be a financial vehicle that facilitates a land-use change that delivers more—for biodiversity, for water, for local climate, and for people.

On the precipice of that muddy hill, I thought of my first few days in Panamá in Gamboa, a stunning rainforest reserve not far from the capital city. I had felt like a kid again, exploring trails and stopping every few steps in wonder. I'd watched a long line of leafcutter ants haul huge pieces of plants, feeling admiration for their persistence and teamwork. The ant brigades never stopped marching; they overcame any obstacle in their way. A capybara, a cute rodent the size of a dog, darted across the trail as I ducked below dangling vines. I'd seen my first corotú tree on a walk up "radio hill" in Soberanía National Park, where American Special Forces had trained during the Vietnam War. Under an umbrella of tangled vegetation, I held the corotú's curious seed pod, a natural replica of a monkey's ear, in the palm of my hand.

A healthy forest is a feeling, I'd thought, remembering the sheer magnificence of that feeling as it pulsed throughout my body.

"Okay!" César said decisively, snapping me back from the corotú canopy to our present situation. "Let's go."

He let out the clutch. We dropped down the slippery descent. A box of first aid supplies for the planting crew fell off the seat between us. I held on as we bounced back and forth.

There were a lot of reasons why I'd chosen to visit this project in Panamá, in the grand finale, if you will, of my quest to understand the science behind, legitimacy of, and hard work involved in this global movement to restore forest cover—however one defines *forest*. I'd originally come across it through Earthshot Labs, a venture capital–backed start-up that brings together investors, carbon buyers, and land stewards to finance ecological restoration projects that also benefit local communities.

The company had received $14.3 million in funding from early investment rounds. In its first ten months of operation from 2021 to 2022, Earthshot had grown from two to some forty-five employees who were working to develop one of the largest pipelines of carbon-financed restoration initiatives in the world. On a walk I'd taken in Marin, California, with cofounders Troy Carter and Patrick Leung, Troy had told me they were addressing the supply problem for certified carbon projects that deliver on more than carbon sequestration. They were doing so mainly in forest ecosystems. He used an analogy from Airbnb, the online platform that connects people who want to rent out their homes with people looking for accommodation. The "soft side" of the Airbnb market, as he explained it, is providing the ability for someone to find a place to stay, click on it, and buy it as if booking a hotel room. He was drawing connections to the increased demand for forest carbon initiatives and other efforts to protect and restore nature. The "hard side" of the market involves a lot more problem solving on the back end to bring good homes online and put standards in place.[2]

"We work on the hard side for restoration," Troy had told me, building out the supply to meet rising demand. "A corporation can say, 'Okay, here's $100 million.' Well, that $100 million means a decade of work for thousands of people, and it will always be very complex. What we're saying is most of our work will need to be on the supply side, the hard side of the market, where real people and real trees need all the tools, support, and money that they can get to do their job. So, all our work has been about supporting land stewards to help scale up restoration activities." By early 2023, around the time I went to Panamá, Earthshot had shifted to a fee-for-service model after growing too quickly with the start-up money. I wondered if investors thought the forest carbon market was too uncertain and the rate of return might be too slow compared to other opportunities. Troy said the core of their work remained the same: catalyzing project development and investment.

Andrew Coates, my local guide and the person leading operations for the project, had been Earthshot's fourth employee, but he'd recently left and was forming a new company, Ponterra, with three other former staff members. Its focus would be on restoring degraded and marginal agricultural land with landowners, and the effort in Panamá would be a pilot for other tropical regions.

I figured if Earthshot was involved in the reforestation effort on the Azuero Peninsula, there was another company already waiting for the carbon credits, and it was probably willing to pay more for a project that went above and beyond carbon. Within the first hour of meeting Andrew and Malcolm from One Tree Planted in person, I'd learned that the company was Microsoft, or *could* be Microsoft. Andrew and the other project developers were trying to finalize a deal for $14 million to secure about half of the credits and possibly more later. It was far from a race to the bottom on pricing. Microsoft wanted high quality, and it was in very early discussions about $49 per tonne.[3] I understood the final price could evolve in the process, but they were focusing on three to four times the average market value for forest carbon at the time. (Later that summer Andrew told me they had three other companies bidding, clearly showing the demand.) If the agreement closed, it would be a prebuy scenario—payment up front for what they hoped and expected the new forests would provide later.

The project had intrigued me for other reasons too. It would create 300 to 400 jobs in the first five years. Then, moving forward over a thirty-five-year period, it would sustain a couple hundred jobs or more. Landowners could double their income by moving from raising cattle to growing trees. Women and children were helping collect seed; members of the local community had been cultivating and planting nearly eighty different species already.

I'd also read an article that ran in the *New York Times* about Pro Eco Azuero, a local organization dedicated, since 2010, to habitat restoration, sustainable land management, and environmental education in the region. Its aim has been to help create a wildlife corridor—spanning seventy-five miles across the peninsula—by reconnecting the islands of fragmented forests. The reporter had spent six "grueling" days in pursuit of the endangered spider monkey, a species requiring larger areas of forest to sustain its population than currently exist. Then, on my very first walk on a property that had been protecting existing forest patches and reforesting other areas, I watched a family of those rare monkeys swing through the branches above me, their long and skinny limbs propelling their bodies forward with grace. A pair of white-faced capuchins leaped from one tree to another to join them.[4]

What had attracted me most to the peninsula, however, was the potential scale of the project and the approach to achieving it; the early descriptions I'd reviewed reported 10,000 hectares (an area about three-fifths the size of Washington, DC) with the potential for expansion to 100,000 hectares. Reaching 10,000 hectares and more could only be achieved by farmers with relatively small or medium-size properties coming together, uniting their collective lands to build and protect contiguous forest again. Plans also included assessing changes in biodiversity—not just plants but birds and mammals too. Those outcomes will depend, in part, on the source populations of species in the existing forest patches and whether or not they move into the new habitat in the years to come.[5]

"The worldview we come from is that nature has a right to exist for its own sake," said Troy, referring to Earthshot and its staff. "Nature is not just for utility. This isn't just about carbon dioxide being sequestered. It's about addressing the root causes of the climate crisis rather than just responding to the consequences. It's about regenerative livelihoods and more sustainable relationships to the land, the deeper issues that we need to address as a civilization. There aren't many companies that think about it like this, but there will be." There should be.

I loved his outlook. *Not only companies*, I thought, *but more and more people and whole communities too.*

In my early impressions, the Panamá project had appeared "boutique" in the sense that it was very much tailored to the local conditions—to the history of the land and its current state, to what the people living there wanted, needed, and valued too. But I suspected there was something about how people were coming together to scale reforestation on the Azuero Peninsula that was, in fact, more generalizable or transferable elsewhere.

We didn't make it to Hector's finca. That first drop turned into a real roller-coaster ride, which we took in slow motion, with each new turn unveiling yet another questionable climb or descent. In anticipation of the upcoming push to get 250,000 saplings in the ground during the wettest time of year, Andrew had considered using military trucks. In a year's time, they were aiming to plant about a million saplings. "I like repurposing from the military for something good," he'd told me, "but anyone can fix a Land Cruiser. Stick with those; nothing less." Tires were our weak point, and Andrew's rig needed an upgrade.

On a steep ascent, the Land Cruiser got stuck behind us when it slipped backward and one tire slid into a deep trench. We tried to pull it out with a winch anchored to a sturdy tree, but the winch wasn't working. So, we went to César's truck next. Malcolm had taken over the driver's position in the Land Cruiser. Ross, Andrew, and I set up the tow.

What do plants need to grow? I imagined the stones beneath my feet again. The question kept playing over in my head, as I ran up and down the hill to relay messages between everyone. The most unexpected, honest answer I could have come up with for my son was "This! See this!"

I leaned forward to make hiking up the steep slope easier as I dragged the tow line to Andrew. Our shirts were soaked in sweat at this point.

Trusting his British humor would recognize the absurdity and larger relevance of this everyday situation, I hollered to Andrew, "So, do you think the Microsoft lady would have some questions now?"

He grinned and chuckled with me. "She would have more questions!"

As part of their due diligence process for considering up-front investment, Microsoft had sent a representative from Carbon Direct, a firm that helps advise companies and organizations on how to reduce, remove, and monitor their emissions. They'd also contracted a forester from a local university and a translator to facilitate conversation with the local people involved. Andrew said the forester was quite traditional and noted that, broadly speaking, there are a lot of experts with more knowledge about planting a limited number of species than about running a remote native forest restoration project. "They were a bit soft," Andrew said; he'd wanted and expected harder questioning to reveal any potential snags in the thirty-five-year plan to grow trees. They'd come at the end of the dry season when the dirt roads were no big deal.

"What should they have asked?" I'd inquired, coming across as a bit irreverent. His answer: "How are you guys going to get to all these sites when the rains come and you need to plant?"

I handed him the end of the tow rope for César's hitch and imagined the pressure of getting to the finca if we had been hauling truck beds full of saplings to plant or food for the crew. Neither César nor Andrew seemed fazed; nor did Malcolm, who had seen more hurdles in various reforestation efforts than any of us.

"Right tools, right work, right team, and more saplings. We'll do it," César had said early in our drive, as we'd discussed what it would take to scale up quickly and effectively with native species. His version of "right trees in the right places" covered more: bringing together the right team by starting with local people and their relationships to the land. They'd troubleshoot whatever challenges surfaced across the pipeline—from seed to sapling to planted tree and then with longer-term nurturing, monitoring, and protection.[6]

"We gotta get out of here," I said, realizing I was standing between two trucks that were now tied together. We dashed up the hill out of the way, as César dragged the Land Cruiser forward. A toad the size of a bowling ball leaped onto the bank between Andrew and me, then disappeared into the grass. "Whaaaat?" I yelled in disbelief. "Bufo," he hollered over the engines, referring to the giant neotropical cane toad *Rhinella marina*.

The Land Cruiser was now free, straddling the trench with all four wheels on either side. César turned around at the top of the hill, and the guys nodded in agreement that it was time to call it. Andrew drove backward, his only option until he could find a place to turn around in the deep, narrow tunnel of the road. The widest spot required backing up the bank and making one sharp turn to avoid sliding into a ditch.

"Go brave and fast," César cheered, as Andrew made the risky maneuver. His words caught my attention as a motto for the project or for larger aspirations about restoring forest ecosystems around the world. It's one thing to want as much carbon sequestration as possible from a freshly planted forest; that work is one version of going brave and fast. But the narrow focus on speedy carbon sequestration is also disconcerting. Tom Crowther had made a similar argument in a 2022 *Time* piece: financial systems that value any single part of nature put the others at risk. Tom says that we "need economic systems that value the complexity of nature and the human well-being." I could see the effort for holistic restoration on the peninsula as a very different and more hopeful version of going brave and fast. I hadn't recognized any of the species on their list as among the 100 or so most commonly planted ones. This, too, was another kind of trial.[7]

César said he would return to the crew the next day. There were more than two dozen planters at the finca already, along with two cooks. They would spend twelve days planting before coming out for a break. He referred to Hector, the finca owner, as Don Hector and called him a "trust man."

"He is an important person in the area, in the community. We say in Spanish, *una sola madera*—like a good wood. He is an *hombre de palabra*, a man of his word." César said landowners were watching and waiting to see what would happen with Don Hector's land and whether he'd get paid.

I hadn't realized until that drive together that Don Hector was a day away from signing the finalized contract for his forests in the making and that I would be there for the moment. Keeping the trees growing and standing would bring him a reliable annual income for decades to come.

César said he was pretty sure that about 80 percent of the farmers in the region would join the project. Given its longtime efforts in outreach and education in the community, the organization Pro Eco Azuero had gathered a list of owners who wanted to reforest. There were about 500 interested by the time Earthshot came along. ("It's the perfect setup!" I'd exclaimed when I heard that. "Oh, hey, we just have hundreds of farmers who want to reforest, and we've been building relationships with them for years!") The foundation of local support was there already; Andrew and César had lined up about thirty landowners with relatively larger properties; they'd need to reach about eighty to deliver on a 10,000-hectare conglomerate.

"We found the leaders and respected people in the community," Andrew told me back in town when I probed further. "Once Don Hector gets his first payment, the neighbors will all sign on."

He raised one hand, cupped, with the fingers pointed upward and touching, a gesture signifying one property or one bold leader in the center with many others surrounding in community. He slowly let his fingers spread open, like a blossoming flower for the forest's becoming.

In the Land Cruiser and over various meals together, Malcolm, Ross, Andrew, and I fell into consistent chatter about the project details and the global movement. We talked about using drones for monitoring; there's been so much progress in the technology available since Virginia Norwood's satellites or even the Hansen Dataset. I asked how they were deciding what to plant on the peninsula and where. "We design everything with the farmer," Andrew explained. "We ask his grandparents, if they're still there, what used to be there. We ask the farmer what he likes, and usually he says, 'Oh, we had some of those there and these over there.' Then we consider the future climate and what can survive in more extreme conditions. But we know we'll also change the local climate by building a forest. It's so bloody hot out there on the farms, but you walk into a forest, and it's like someone turned the air conditioning on." We discussed how backyard nurseries at people's homes could yield more healthy saplings and benefits to community members. I thought about Calder's giant sequoia seeds—*Where are the seeds, seedlings, or saplings coming from? Where are they going?*

I asked whether every project requires a leader or champion of some sort to get it going—an Andrew, if you will. I had asked a similar question of my former advisor at Stanford, Eric Lambin, on our walk in the California chaparral four years prior, but Eric had a view of forest cover reversals at national levels—whole countries that had gone from loss to gain. Malcolm and Ross had witnessed them place by place, community by community. Malcolm said that if it's a relatively young organization and a new effort, you need someone to have a vision and start building out the capacity. "In my experience, the rate of scaling can also be determined by that fearless leader." He said many factors need to align, which agreed with Eric's national perspective, but there also needs to be a community of local people who are ready, willing, wanting, and financially able to support forests again. The plan on the peninsula was to progress rapidly from the small experimental sites they'd used in the project's first year to a couple thousand hectares in the next. That would require scaling up every dimension of operations by a factor of five and holding that steady for years.

"It's not about finding the perfect project," Malcolm noted. "It's about finding the perfect mix of potential and desire to build it out right over many years." There's a range of models for how that happens in practice: through an established local NGO that has years of in-country experience, an international NGO or corporation that has good access to funding but may be more disconnected from the local needs, a government agency, local communities, and Indigenous peoples serving as project leaders, or with the support of a creative individual, like Andrew, who is passionate about a large-scale endeavor and hiring the right people. Any model or combination can work, but many stars still need to align.

"People giving OTP money often ask, 'How soon after you get my donation can tree planting happen?'" Ross said over one of our chat sessions. "They often have this expectation of the timeline, and some will literally call back the next day to ask, 'Are they planted yet?' Few people realize what's behind it all, what this takes." My mind drifted back to the freezer in Canada, where producing millions of saplings had taken about two years.

Despite my extensive research, there were still some issues that hadn't come across my radar until Panamá. Take payments, for example. A lot of organizations and companies work on monthly billing cycles. Grants can take even longer to process. "But people here need to get a first payment as soon as the trees go in the ground," Andrew told us. Delays can damage trust between people who have never worked together like this before and need to work together for years to come. Malcolm said OTP tries to buffer this by supporting a project during the early phases, such as nursery development or site preparation. "Good projects should get to scale quickly while still delivering high quality," he explained. "We need to start running, not walking."

By this point, I'd completed nearly 150 interviews with people around the world who were involved, in one way or another, in fostering forests of the future. Yet my days on the peninsula still felt like a crash course in all that needs to align, again and again, to make growing them and keeping them *maybe* possible. I could see the convergence happening there before me, but how it will all play out still felt—and feels—uncertain to me.

Never in human history have people taken a global approach to planting new forests and ecosystems. There isn't one sure bet for what may work.

When Andrew, Malcolm, and I traipsed along the hillside at another property to investigate the trees they'd planted at the end of the last rainy season, an unexpected feeling of vulnerability surfaced inside me. That same sensation arose later at the central nursery we visited, when I walked the narrow aisles between thousands of little saplings growing in trays. There were about 45,000 there at the time; in two or three weeks, they'd reach 135,000.

Mahogony. Corotú. Guayacán morado, the rosy trumpet tree; its pink, puffy flowers bloom during the dry spells and hang softly like cotton candy in the canopy. Moringa, a highly drought-tolerant species. Ceiba, known for its stunning buttress roots, these wide and wandering planks that extend outward in a maze of support. The tiny ceiba crowns were still enclosed inside seeds the size of coins. Together, hundreds of them looked like a crowd of green stick figures doffing their brown hats.

It took me some time to decipher the oddly familiar sensation.

I felt like the mother I am and the mother I was standing over my baby's crib, reminded that every life is a miracle of nature—the perfect convergence of many factors and intentions. Yet all the love and care in the world can never ensure that every individual leads a long and healthy life. Some forces are outside our control.

But I am still here for you, wanting and wishing for you to grow strong, giving you the best shot I can, and hoping some luck does the rest.

What will become of all these hope-filled forests from this blip in time when more people want them back?

After the failed finca expedition, we stopped in a tiny town called Bayano, where Don Hector's family ran a little shop beside his home. We bought Coca-Colas from his daughter and sucked them down in seconds, returning the glass bottles before going to see the backyard nursery. Don Hector had committed to growing 10,000 saplings that season; his daughter was working toward the same. They'd received training in how to mix the soil and start-up materials from the project: trays and an organic fertilizer, which César made locally and Andrew called "tree Viagra." Each "micro-producer" in the community would get paid forty cents a sapling. Earlier in the year, they'd collected seed from dozens of native species by walking the forest patches that they knew best and seeking out mother trees not far from their homes.

Andrew said the work particularly empowered women in the community. The project would also get more saplings this way. If Don Hector and his daughter both hit 10,000 that year, they'd each make $4,000 in just a few months. "If you just have all the plants in one central location and someone forgets to go water them, then what?" Andrew asked. "But this way, everyone is responsible for their own crop. We come back and buy them when they reach about thirty or forty centimeters."* He mentioned a

* I later shared this model with Alvaro Umaña, as we continued to stay in touch after my initial interviews. He said they'd used a similar approach in Costa Rica in the 1980s and 1990s during the surge of forest protection and restoration activities that ultimately doubled the country's forest cover. "Women, by nature," he said, "are often great multitaskers. This is something they can do from their home, and it gives them an opportunity for their own income." Malcolm said about 50 percent of the projects he'd supported through OTP in the Caribbean and Latin America were doing something similar.

woman named Miriam who could make more money growing saplings at home during those months than her partner could in construction. When we later met outside her home, she told me her twin boys had collected their seed, and the children also helped with watering too. I was reminded of a page from *Planting the Trees of Kenya* with an illustration of children and women tending to garden plots and small saplings. "All this hard work, but the women felt proud," the story goes. "They had work to do, and the work brought them together as one, like the trees growing together on the newly wooded hills." I felt like I was inside the Panamanian version of the children's fairy tale, so moved by the progress and inspired by the people coming together, yet still unsure as to how it would pan out.

Across the peninsula, men and women were growing saplings at home, and others worked at the "mother nursery" too. At this central nursery, they cultivated saplings to bolster capacity, adding supply to what the microproducers could yield. Beside the Rio Orio, which supplied water, "La Madre" would also serve as a temporary home for saplings collected from Miriam, Don Hector, and many others before planting. During the peak of the preparations, some nursery workers slept in a repurposed US Army tent on site. There was an old shipping container at the mother nursery that they also planned to turn into a seed bank—"Not the $100K version like Terraformation's," Andrew clarified, "but $2K style."

In Don Hector's backyard nursery, I also saw tiny fruit trees and coffee saplings. He'd plant them on his land for another source of income and sustenance. Andrew told me that the main limiting factor for scale at the time I visited was sapling trays. *Imagine that! Trays!* I thought, caught off guard. Those little plastic trays reminded me of the Belgian scientist Patrick Meyfroidt's thoughts on taking any endeavor to a global scale: try going big, and unexpected issues often arise in the process. We just deal with those next.

"We are scouring the earth for trays!" Andrew said. They were working with a local company to try making some from recycled plastic and were waiting for new ones to arrive from Colombia. They'd paid $21,000 for a shipping container of trays, the highest-priced item per tree planted; in the years to come, supplying the project would require eight or nine more shipping containers of them or great success with the recycled prototype.

This is scale in the making, I thought—a mother nursery with plans to reach capacity for a million saplings; some thirteen different communities with individuals committed to producing saplings in their backyards; thousands of trays; soil mixed with cattle manure, sand, molasses, various microorganisms, and ash from wood collected locally to add potassium, calcium, and other nutrients; dozens of farmers ready to reforest their land.

"It takes people with good hands and patience," César told me of any beginning. "Going from seed to plants big enough for the land: those three months are like caring for babies."

We slept each night in cabanas by the ocean and rose early. Time flew too fast; I wanted to stay. I wanted to help. I wished my son were with me too. I wanted him to hear the howler monkeys and walk the aisles of saplings at the mother nursery. We would admire the peculiar tropical species together: that one with triangular protrusions jutting from its trunk like shark's teeth; the corotú with its monkey ears. I wanted him to meet Miriam's boys or any other children who had collected seed. I wanted him to see and feel a community of people working to reconnect the fragmented native forests. I saw those children as being present in their own "baseline," in the wonders of the natural world even in its current state—what I see as degraded and they might see as beautiful. But they are still striving for something better.

Renovate *(verb).*
To impart new vigor.
To remodel.
To revive.
To repair and improve something.

Ourselves, our families, our communities, our species included.

On my last full day on the peninsula, we went to visit Pro Eco Azuero, the organization that had been working in the region for over a decade.

I expected offices, and there were a couple. But it felt more like a small school campus, offering the kind of hands-on activities and environmental classes I'd want Calder to experience. There was an outdoor classroom with chairs lined up across a checkered floor. Children ran around the yard, and they helped transplant saplings. One section of the property had tarps hanging above their crop like I'd seen at Don Hector's and the mother nursery.

The children played a game of musical chairs, but the chairs, as they learned, represented trees that provided habitat for many species. I watched as the ecological dynamics played out between the children; they became more competitive as their teachers removed trees, one at a time. They raced each other to claim a home when the music stopped. When someone planted more trees, bringing chairs back, they danced together in celebration as the strain lifted.

"We are training them to be environmental heroes," Sandra Vásquez de Zambrano, the executive director, had explained to me in her office. They were engaging about 100 children in the region; some students came to them, but the teachers also visited rural schools throughout the corridor. Undoubtedly these children go home to their parents, many of whom are farmers, and share their own versions of what treekeeping can change.

While I observed the class, Andrew was in a back office, printing the final version of the contract for Don Hector. I watched the children gather to play a game that looked like a forested version of Chutes and Ladders. Trees extended in various directions across the grid of numbered squares; some trees stood tall with full crowns; others without leaves appeared cut or toppled.

Roll the dice and get five, for example, and I'd land on a square that said (in Spanish), "A farmer planted a tree just beside his creek." Then I could advance to another square by following the tree to its crown. Other boxes described loss and unrealized potential: a tree cut or another not planted yet. Fallen trees took the players back down the board in these scenarios.

"What do you call it?" I asked Sandra in her office, inquiring further about the game.

"El Bosque de la Suerte," she said. "The Luck Forest."

"That probably isn't the best translation; there might be a better name in English," Malcolm commented.

"Trees and Snakes?" Ross brainstormed.

"I think Luck Forest is pretty good," I interjected, surprising myself with my own conviction.

Through all the research I'd done, the scientist in me would have liked to reveal the perfect recipe for How to Reverse Forest Cover Loss to Gain and Keep It. Heck, the reporter in me has sought the same—for carbon sequestration and much more. It just doesn't seem fair that some luck will be needed on the peninsula, or anywhere else for that matter, given all the hard work, planning, and good intention.

I found myself thinking that many of those saplings I had stood beside at the mother nursery would probably become trees. And together those trees would become some version of a forest, delivering, at least partially, what people and other forms of life need to survive. But that forest was still just a possibility, an intention, waiting to be realized with the perfect convergence of everything trees need to grow over the passage of time. It's not very scientific or personally empowering, but my honest answer to "What do plants need to grow?" includes a bit of luck.

I sat beside Don Hector in a little café as he and Andrew reviewed the contract one more time. Don Hector wore a *sombrero de junco*, a classic Panamanian hat. His posture was impeccable.

"Listo? Quiere firmar?" Andrew asked. *Ready? Do you want to sign?*

They nodded at each other and then elegantly penned their names on two copies of the first agreement for planting trees and the sale of carbon credits on the Azuero Peninsula. I got goosebumps, and not because I thought that the potential carbon sequestered would significantly alter the current trajectory of the global climate system. I got goosebumps because I felt confident that the money transferred to that land and its people for the hopeful creation and stewardship of future forests would restore life on the peninsula in ways that extend beyond the value of any singular benefit or "service."

"Cómo se siente?" I asked Don Hector. *How do you feel?*

"Estoy emocionado," he said, "porque vamos a reforestar." *It's emotional because we are going to reforest.*

He wanted and needed to transform the land. He is not alone.

I recalled one of the last questions I'd asked Sandra at Pro Eco Azuero: "What do you see unfolding with this new project?" I had been probing about the goal to achieve scale with native species and local leadership and participation. "What's your vision for what it will become?" I added.

"If we want to see change in the region," she replied, "we have to reforest at least 10,000 hectares. That's the only way that we will really make a difference." If the effort unfolded too slowly, she thought it would be hard to bring their vision for a more forested future to fruition.

"We need all the help we can get," she added.

Go brave and fast, I thought, knowing that the beginning matters, but so does the keeping across time. I still believe it's worth striving to create a better baseline that comes next.

16

WITH DREAMS OF FUTURE LIFE

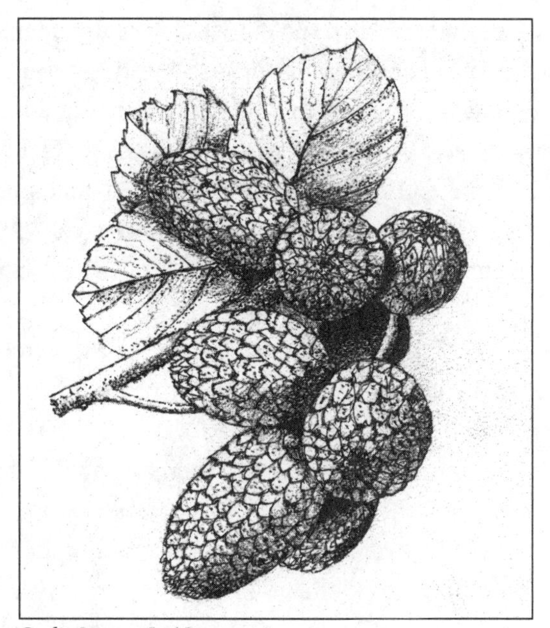

Credit: Lorenzo José Rosenzweig

"Are we doing the right thing?" Andrew Coates asked. I wasn't sure if he was talking about the restoration on the Azuero Peninsula or the global movement, and the question was rhetorical anyway. *This is someone trying to rebuild native forest and integrate more trees with other land use too*, I thought. *The effort is about creating habitat for species, providing jobs and income for landowners, protecting water resources, regulating local climate, sequestering carbon, collaborating across cultures and communities. The list goes on. The project in Panamá probably checks every box for the United Nations' Sustainable Development Goals. And even Andrew is still questioning what's right?*

"Is this a solution? Is there a better way to do this?" he continued. "There's this enormous movement happening, but nobody seems to really know what they're doing, truly, because it hasn't been done before. We've had ridiculous situations where hundreds of thousands of the wrong kind of tree were planted in monocultures across the world. That approach had consequences, but people still supported it and did more of it. Is *this* one good?"

He said it's important that everyone involved keeps asking and discussing, "Are we doing the right thing?"

I feel very fortunate to have spoken with so many people around the world who are wanting and hoping for and working toward a more forested future. At any other point in history, our conversations would not have been possible. The collision of the climate crisis, the boom of planting interest, and the COVID-19 pandemic meant that people whom I never could have met in person were willing to meet virtually. Many dedicated people have informed what I now think of as "better" moving forward.

"It's like everyone involved is striving to be a treekeeper to some extent," I told Andrew. "But no one knows for sure how each effort will pan out or what they'll achieve collectively."

Maybe it's less about right and wrong and more about striving for better as people learn from the challenges and successes and what unfolds in our forests, new and old, over time. Business as usual supports a trajectory toward the demise of life as we have known it—both within our forests and outside them. We can only hope that as many people as possible work toward charting a different course.

Looking back over the past decade, I see how the narrative has evolved. Susan Cook-Patton at The Nature Conservancy summarized the progression best: "At first people had to accept that climate change was a real problem. Most people are convinced of that now, hopefully. Then when it came to natural climate solutions, there was this 'either/or' debate, like natural climate solutions were dangerous distractions to the other work people needed to be doing. But no, it's 'both/and'!" The scientific community wasn't saying that people need to tackle natural climate solutions instead of fossil fuels. The message got distorted. Susan continued,

> Now, I feel like that either/or debate has mostly been laid to rest. Then it seemed like there was a focus on a "sole solution," like, "We must focus exclusively on tree plantings!" But whoa, hold up, there's lots of options for how to get more trees in the ground! And by the way, protecting native forests is usually much more effective than trying to restore them afterward, while embracing the diversity of options out there in attempts to do so. Now the narrative needs to move to "How do people do this work on the ground in ways that are sustainable and will last and positively impact all the people that live in those places?"

We've come a long way from more people accepting that climate change is a problem to wanting to do something about it by sequestering more carbon in forests and reducing, removing, and avoiding emissions through a variety of approaches. Nevertheless, I think the narrow focus on trees for carbon detracts from the much bigger picture of biodiversity for well-being. For millennia, Indigenous people have lived with a holistic understanding of the interdependence among all life forms, and they founded cultures and communities on relationships of mutual care. Tom Crowther told me that everyone involved needs to reframe the global movement by focusing on biodiversity for well-being, and the work should be supported by millions of local efforts.* Trees are, of course, part of the world's biodiversity, and carbon is also tied to well-being—not just *human* well-being but the well-being of all life. Yet acting in the interest of biodiversity for well-being means anyone and everyone can contribute in one way or another to restoring nature, including forests. A renewal of this kind of thinking and valuing of nature encourages more people to do the hard work of sustaining its functioning for more benefits than one.

The dream is to make acting on the instinct to protect and care for nature possible in the world we've created today. That would be a revolution because there's no way back to yore.

* As of January 2024, Restor, the online hub for restoration and conservation efforts that Tom helped start, had about 148,000 projects in its database. The numbers keep growing every day, and undoubtedly there are many other efforts underway that still fall outside its radar.

Marc Benioff, the CEO of Salesforce and a speaker at the one-trillion-trees press conference in 2020, was an early promoter of planting a trillion trees and tracking that progress by tree counts. But the focus at Salesforce has evolved as well, hopefully serving as a model for other companies in the years to come. One of the last people I interviewed was John-O Niles. At the time, he was the senior manager of natural climate solutions at Salesforce. John-O is an international expert in forestry and climate change policy. He also wrote the first global standards that demonstrate a project is not only addressing climate change but also supporting local communities and conserving biodiversity.

"When you're looking specifically at increasing forest cover, is any approach okay?" I asked John-O over a call. "How about a fast-growing *Paulownia* plantation or a native restoration project with some ninety different species?"

John-O said we desperately need more money dedicated to supporting good projects, whatever they are. Plantations can alleviate pressures to use native species, and they provide materials that people will always need. Even so, he admitted, "Salesforce probably won't get behind those plantations because of the optics of supporting a monoculture." Plantations in some places may be good for the world, but the company prefers to support native-species restoration and projects with mixed species and to focus more broadly "on ecosystem services—not just the trees."

I know that five-star restoration, which fully restores ecosystems, can't go everywhere that is possibly suitable for trees or a forest. There's too much demand on our land. Nevertheless, I do think any effort to increase forest cover can start by considering how far a project could go toward achieving more holistic restoration.

My husband says the term *ecosystem services* suffers from terrible naming. I understand the critique, but I also disagree. To me as an ecologist, it captures so much of the complexity in the relationships between people and nature in just a couple words. It implies an exchange and mutual care. If people want the services that nature provides, they need to pay for and provide the conditions that enable it to thrive. The increasing drive to get more people, companies, and governments to invest in the future of the world's forests is not just motivated by a love of trees. This is about survival.

Just as Matt struggles with the unwieldy ecosystem services label, the public has had a hard time understanding the "biodiversity crisis" or the importance and meaning of the word *biodiversity.* I typically turn to ecologist Aldo Leopold to explain the importance of biodiversity: "To keep every cog and wheel is the first precaution of intelligent tinkering." But we're not going to keep every part. There's a ton of overwhelming data indicating a trajectory toward the next mass extinction. Every day, we're still losing species and football fields' worth of tropical forest. Aiming high to keep the cogs and wheels pushes each endeavor toward the five-star restoration standard, offering more long-term possibility and promise. We don't need to choose between addressing the climate crisis or the biodiversity crisis. This restorative work is the way to tackle both.[1]

"There are a lot of 2030 targets," I said to Tom Crowther during a follow-up interview in June 2023. "But what do you want to see happen halfway through this decade? What do you think is possible?"

"I want the biodiversity movement to be as big as the climate movement," he replied, "and I want no organization to claim that they're doing anything good for nature without showing it."

In the fall of 2023, philanthropist and former Microsoft CEO Bill Gates told a *New York Times* reporter that he didn't use "some of the less proven approaches" to offset his own carbon footprint. This occurred at a climate event in New York City. I was home in Montana, finishing my book research while the aspens turned yellow in autumn. Gates mentioned funding heat pumps, which are energy efficient, and installing solar panels as examples of where he does attempt to offset. He said, "I don't plant trees." Several colleagues sent me the article, expressing frustration with the continuance of simple statements that can do a lot of damage.[2]

Jad Daley, the CEO at American Forests, was also quoted in the article. I contacted him, and we exchanged emails. I wrote, "I partly wondered if [Gates] was being intentionally provocative to shift the dialogue a bit with respect to carbon markets—trying to push more people to reduce emissions or offset via other mechanisms. There must be a limit to how much of the forest work becomes credited and offset, right? And if so, that means more leaders need to step up to fund the work for more than net zero goals." I was referring to the need for people to support forest protection and restoration to harness more benefits than just the potential for sequestering carbon. Jad suggested that Gates and other critics of forest carbon credits still see value in these natural solutions to climate change. Marc Benioff at Salesforce even pointed out to the *Times* reporter that he and Gates had both invested in a company that sources carbon credits for restoring nature. Jad argued that people need to stop conflating the questions "Can forests help us sequester more carbon?" and "Are carbon offsets a legitimate way to support the endeavor?"

Plant/don't plant, I thought. *This approach is the best/definitely not that one. Offset with forests/don't offset with forests. Is it all so black and white?*

Jad had readily acknowledged that even a trillion trees couldn't, on their own, absorb what the reporter had called "titanic amounts of planet warming emissions humans continue pumping into the atmosphere." "Forests alone cannot solve climate change. Not by a long shot," Jad told the *Times*. "The conversation has become cartoonishly simplified. It doesn't need to be that dualistic."

People tend to want simple answers; that's one reason why the media thrives on flashy headlines. When I started my research for this book over four years ago, I was thinking about the question "To what extent can trees really save us? And how?" I, too, was focused on the climate crisis. I didn't think finding the answer would be easy, and I saw value in considering the complexities. In our emails about the article, Jad wrote that a less "cartoonish" conversation would focus on the current and future potential of trees and forests to sequester carbon, what people must do to capture that potential, and how we pay for the work required. Through my journey, my inquiry evolved as well. It became about reconciling the competing viewpoints regarding the potential of forests to reduce greenhouse gas emissions and discovering how to grow more forests in ways that deliver more benefits over time. My answers aren't as simple as headlines like "Plant Trees" or "Stop Planting Trees," but this is where I land.

ON POTENTIAL

It feels like rewinding the narrative to focus on "potential" only with respect to carbon sequestration. So many other benefits come from keeping our forests and growing more. A 2021 study, published in the journal *Nature Climate Change*, estimated that global forests sequestered 7.6 gigatonnes of carbon dioxide, equivalent to about 2.1 gigatonnes of carbon, per year from 2001 to 2019. By its calculations, this "net carbon sink" comes from the difference between what our global forests *sequester* on an annual basis (15.6 gigatonnes of carbon dioxide, or 4.3 gigatonnes of carbon) and what they end up *emitting* (8.1 gigatonnes of carbon dioxide, or 2.2 gigatonnes of carbon) due to deforestation and other disturbances that occur naturally and are caused by humans. Scientists involved with the Global Carbon Project consider this "flux" specifically in terms of changes cause by humans; the estimate reported in the 2023 global carbon budget for the average amount of carbon dioxide sequestered annually on land from 2013 to 2022 is lower than the estimate from the *Nature Climate Change* study. It's 3.3 gigatonnes of carbon per year (12.1 gigatonnes of carbon dioxide). So, based on that number, forests and other terrestrial ecosystems are capturing almost one-third of our global CO_2 emissions in recent years. Let me reiterate: this is what our existing forests are *already* doing.[3]

There are country-level and regional estimates of this *current* service that forests provide, and others calculated by land ownership. From 2001 to 2021, for example, Indigenous forests across the nine Amazonian countries removed about four times as much carbon as they emitted, making them net sinks of 0.34 gigatonnes of carbon per year—about equivalent to the United Kingdom's annual carbon emissions. Forests outside the Amazon's Indigenous lands were a source of carbon emissions due to more rampant forest loss.[4]

I take these estimates as proof positive that forests are already providing such an enormous service to us all by removing carbon from our atmosphere. Further degradation results not only in troublesome emissions but also in the loss of many other critical services. However, there is still the *unrealized* potential from forests, meaning what growing *more* trees could offer.

On November 13, 2023, more than 200 scientists finally reached consensus—around 226 gigatonnes—on the total natural potential of forests to store more carbon at a global scale. The study, published in *Nature*, was called "Integrated Global Assessment of the Natural Forest Carbon Potential," and it made the *New York Times* and other top media outlets that same day. The *Times* lead promoted the estimate as equivalent to about a third of the amount that humans have released since the beginning of the industrial era—what Jean-François Bastin and his colleagues had called the "carbon burden" in the 2019 Global Potential study. The abstract in *Nature* stated, "At present, global forest carbon storage is markedly under the natural potential."[5]

Tom Crowther, the study's senior author, told the *Times*, "If we continue emitting carbon, as we've done to date, then droughts and fires and other extreme events will continue to threaten the scale of the global forest system, further limiting its potential to contribute." He was careful in his framing—clearly emphasizing the need for more *treekeeping* as opposed to *tree planting* that had hooked so many people, companies, and governments.

The scientists involved in the *Nature* study had estimated potential using different modeling methods. I took it as a "come one, come all" approach to bring the top scientists together, let everyone answer the same question in various ways, and see what happened. Tom had shown me a figure during a video meeting in June 2023, while the article was in the final stages of review. Different model types were on the *x* axis. Potential carbon stock was on the *y* axis. The columns for potential carbon stock from each model were all around the same height, ranging from about 200 to 250 gigatonnes.

"This is almost laugh-worthy," I exclaimed, reflecting on the attacks on Jean-François Bastin and the Global Potential study's estimate, just four years prior, of 200 gigatonnes of carbon that more forests could sequester. When *Nature* published the new study and I scanned the list of authors, I immediately recognized some of the most vocal critics of the Global Potential analysis. It felt like "ah, one big happy family" and a long-overdue apology to Jean-François.

It also offered a new opportunity to better direct the movement that had focused primarily on planting because the researchers considered how much of the potential for carbon storage comes from conservation of existing forests as opposed to restoration. The Bastin analysis hadn't provided that breakdown. In this hypothetical situation of a world without people, the Global Potential study of 2019 had assessed the *natural recovery*—what would occur on nature's own time scale and trajectory.

With a lot of people subsequently arguing about *how* to harness that natural potential and debating whether it should come from new forests or old forests, such a distinction really matters now. It could help inform what actions people will take and where.

"Imagine that we've removed half the world's forests and we've degraded what's left in an oversimplified world, which is not entirely reality," Tom posited during our call. He was bringing me back to the three-trillion-tree count—half as many trees as when civilization arose. He was also making a point about ecological integrity, meaning how well our forests are doing and the extent to which humans have modified their state already. "So, the remainder of the forest that is still here is fragmented and degraded and secondary growth or tiny." He said the average standing forest today is at about 30 percent of its full maturity.

"The coolest thing we learned is that 60 percent of the potential is just in the areas that are *already* forests, allowing them to reach their maturity or helping them do so," Tom declared. The other 40 percent could come from creating more forests or, as he clarified, "recovering biodiversity in degraded landscapes."

I appreciated both the consensus and the breakdown, but I kept thinking about *real* potential—what's socially, economically, culturally possible. Tom referred to the 200 or 226 gigatonne results as "where forests would be naturally—without us" but still outside croplands and urban regions. Bastin had never made the *future* projections that anyone trying to solve the climate crisis with various strategies wants; the researchers were saying what could naturally occur in a hypothetical world without people.

It should be clear to more people now that "a trillion trees" isn't a trillion *new* trees. It is keeping trees, growing trees, planting trees, letting them regenerate on their own, and creating the conditions to support their natural recovery.

The rate at which carbon could be sequestered is another issue; it's impossible to achieve 200 gigatonnes for 2030 targets. "I feel like we are on the cusp of this incredible future, but we need more time to achieve it," Susan Cook-Patton told me when we talked about the potential of restoring forests. "Achieving restoration at scale means talking to a lot of people, figuring out what they want from their landscape, building nurseries that will be full of native tree-planting stock, as well as incorporating local communities into the design, implementation, and benefit sharing of the restoration activities. We need more time."

It's also important to recognize that carbon emissions are only part of the climate change problem. Methane, for example, has caused about a third of the warming from greenhouse gas emissions since the Industrial Revolution; carbon dioxide is responsible for about half. Even if people succeed in reducing carbon emissions and sucking more out of the atmosphere, there are other gases to address.[6]

Several studies have nevertheless attempted to project how forests could contribute on an annual basis to the global effort to reduce, remove, and avoid emissions. It takes a little wading through any one of them to discern the role of forests among other natural contributors, such as peatland restoration or improved grazing practices. Each study also groups forest practices in a different way; some separate reforestation from ecosystem restoration, and others distinguish reforestation from natural forest management and forest fire management, for example. Yet the projected rates for annual sequestration (using a 2030 target year) fall similarly around 1.2 gigatonnes of carbon or 4.5 gigatonnes of carbon dioxide per year when combining the categories that indicate increases in forest cover. Studies take different approaches to factoring in the cost or economic feasibility, by accounting for the social cost of pollution, for example, or the foregone opportunity of other land use. A 2021 report by the Coalition for Negative Emissions, supported by McKinsey & Company, accounted for the economic feasibility of growing more trees by considering where the alternative land use, such as agriculture, would be more attractive economically. Accordingly, it offers a lower annual estimate of potential reductions that could come from reforestation, natural forest management, and agroforestry (trees in cropland): 1.6 gigatonnes of carbon dioxide per year.[7]

That's so much less than what nature might do in the world without us, I thought when reviewing the numbers, *but it's still something.* We need every solution possible.

Tom Crowther thought results from the Coalition for Negative Emissions were "reasonable on an order of magnitude" but had questions about the methods for the economic assessment. He also said that he is far more confident modeling the amount of carbon stored in forests globally than in projecting future sequestration at any one location. If we consider restoration as a process of finding solutions that make forest regeneration the economic preference for local communities, then choosing to use land for trees becomes much easier. Then we could harness far more of the potential of forests to sequester carbon and deliver other benefits too. Sustaining these restored forests over time requires more funding than currently offered by the lowest prices for carbon.

To make sense of it all, I look at the estimates of how much *more* we could sequester with more trees—1.6 to 4.5 gigatonnes of carbon dioxide—in relation to a recent estimate for carbon sequestration *already* occurring on land: 12.1 gigatonnes of carbon dioxide, or about 3.3 gigatonnes of carbon. That number comes from the 2023 global carbon budget, which details the amount of carbon dioxide that can be "spent" (emitted) for a given level of warming. The comparison suggests that more trees could increase current land-based sequestration by 13 to 37 percent. Consistent with this, Tom Crowther says we could increase that service by about one-third annually. But another way to look at the range from 1.6 to 4.5 gigatonnes of carbon dioxide is to consider it alongside the world's total annual CO_2 emissions in recent years: about 37 gigatonnes.[8]

My main takeaway, if anyone wanted to make a headline out of it, is that growing more forests can make a meaningful contribution to the fight against climate change by removing carbon from the atmosphere, but trees are far from the silver bullet that people had hoped.

A forest is a community, supporting and protecting life near and far. A forest is a calming force for our minds and bodies. It is a feeling, beckoning our children. We need more forests, and we need to keep the ones we have. Capturing the potential of their carbon service and beyond, in my eyes, is about "better." How do we better do what people are already doing and get as close as possible to five-star restoration? Malcolm Porteus González from One Tree Planted says the community needs to want the forest; the community needs to care for it; it needs to be economically viable. That's a strong start.

Many people have put a lot of time, effort, and money into quantifying carbon in forests, and those methods and estimates are still improving. Yet, if we want to aim higher, improve our monitoring of the outcomes from these restoration efforts over time, and sustain more trees, getting better at quantification and valuation of many other services is also important. "People want simple forest recovery metrics that are monitored remotely," ecologist Daisy Dent, a long-term research associate at the Smithsonian Tropical Research Institute, told me. Those can be costly and require technical skills. "We need to have more options and especially methods that empower more local people too."

"Better" should include improvement in the tracking of commitments and transparency. Can more people be transparent about how every project plays out over time, so that others can learn from the failures and successes?

Other payment-for-ecosystem-services schemes and credit systems, in addition to the carbon ones, are becoming available. Biodiversity credits, water credits, nutrient credits—these are some. Buyers of such credits could promote the benefits they are facilitating through careful tracking of a project's outcomes, and the funding would help pay for the conservation or restoration work at a given site. I hope these emerging tools encourage more widespread positive action and avoid exacerbating inequities or allowing parties to continue degrading nature while others pick up the slack.

This work requires trust and collaboration across sectors, communities, and cultures like never before in history. The intention behind quantifying and economically valuing nature should not be to profit from it but rather to fix the economic system that has overlooked, for far too long, how functioning ecosystems sustain life. The "potential" is about a lot more than carbon sequestration, and if asked for my opinion, I'd remind any prospective funder to start with a more holistic view.

Looking back, Paul Faeth said the early forest offset projects supported by the global energy company AES had a similar appeal: "Even if you completely discount the carbon, there are good reasons for protecting forests. There are good reasons for growing trees. There are good reasons for sequestering soil carbon. All these things have massive co-benefits, and for that reason, I think the efforts are worthwhile."

I asked John-O Niles at Salesforce about the evolution from "number of trees planted" as a marker of success to defining success as something better and more holistic. Even with ongoing efforts to improve monitoring of the outcomes from forest-growing initiatives, that one tree-count metric always seems attractive. "The number of hectares restored or conserved or made more resilient or that are helping improve livelihoods are probably better indicators," he said. "But, for instance, when I talk to my mom and tell her, 'I helped write the standard for a carbon market' or 'I'm engaging the Paris Agreement on Article VI, which sets out how countries can pursue voluntary cooperation to reach their climate targets,' she never really understands what I do. When I tell her my job is to fund a 100 million trees as part of the movement for a trillion, she says, 'Oh John-O, you're doing such good work!'" Environmental narratives are always complicated, even more so when you're talking about an invisible gas that practically everyone generates or about places with high biodiversity that are far away from our urban centers.

John-O added, "We also never assumed this was about just trees. We've always said, 'We don't care about trees; we care about what trees do. At the same time, we will hold ourselves accountable in a way that is really easy to understand."

ON PAYING

Despite the recent surge of funding for future forests, it's not enough. The figure $50 billion per year has been used a lot to describe the finance gap for achieving the Bonn Challenge—350 million hectares of forest restoration by 2030. Some experts I interviewed suggest looking at the gap in funding to address biodiversity loss, more broadly, because restoring forests will help address that crisis too. One recent estimate indicated a funding gap of at least $598 billion, billed annually, for biodiversity loss. The vast majority of financial support for biodiversity currently comes from public funds; the domestic budgets and tax policies of eight countries, for example, contribute about $75 billion to biodiversity per year. Governments are stretched too thin to cover the remaining gaps. We need to find other ways to support the restoration movement.[9]

Many smart people with economic and political expertise have developed strategies for incentivizing the private sector to contribute. Unlike a carbon offset, in which a company turns to a third party to reduce, remove, or avoid emissions, an *inset* involves a company looking to its own supply chain for places where it can change its own practices. It might use savvy marketing strategies to promote its good work for climate and biodiversity, thus attracting more concerned citizens to choose its product instead of others.[10]

Some reports have shown recently that the voluntary carbon market, in which companies or governments opt in to trading credits, may grow to $20 billion to $40 billion by 2030. That's a notable rise from 2021, when it was worth about $2 billion. In 2023, nature-based projects, including forestry and other land-use projects, constituted almost half the market share. There are plenty of problems with carbon credits, but they've funded forest-related work to date and may do more into the future. So, I have been struggling with where I land on them too.[11]

Most people I interviewed suggested that the ways for ensuring high-quality forest carbon projects will improve with time, but some boldly suggested that we should do away with carbon credits that come from forests altogether. Those people are concerned about the projects that overlook biodiversity and local people's livelihoods, the outsourcing of emissions by wealthy people or nations, and their overall effectiveness at long-term carbon storage. I don't think we should do away with forest carbon credits unless we fail to make them better. People should commit to implementing more high-quality projects, while still working to address the legitimate issues in the evolving systems for creating and trading the credits. In a better world, offsetting with forests is only a small fraction of what people do to address the climate crisis. The biggest polluters need to make more emissions reductions, but those emitters also tend to be the ones who worry relatively less about the impacts of a warming world.[12]

During the early trading of certified credits between Costa Rica and other countries, government leaders in Costa Rica and in the United States had discussed offsets only representing about 5 percent of what a country might do to address its emissions. That means only a very small portion of a larger effort to reduce US emissions, for example, would have come from international offsets. In the eyes of some leaders at the time, offsets would be like icing on the cake. Far more reductions would come from sweeping changes within our country. That hasn't happened.

The problem of overreliance on offsets is further exacerbated by the fact that one tonne of emitted carbon isn't 100 percent interchangeable with one tonne of removed carbon. Danny Cullenward, a climate economist and lawyer, calls this issue the original sin of all forest carbon thinking. He told me, "It's the idea that you can measure a stock of carbon, find a tonne in a tree stem or underground, and assume what that tonne does is equivalent to what happens when you burn fossil fuels." His concern, among many, is that relying on imperfect trades of emissions in one place for removals in another will lead to more climate instability later. Scientists call this problem *removal asymmetry*: CO_2 emissions are more effective at raising atmospheric CO_2 than equivalent removals are at lowering it. The increase in emissions puts other changes in motion, so taking CO_2 out after some amount of time has passed doesn't just reverse the chain reaction. The comparison between what's emitted and what's removed also falls short in part because some of the carbon released stays present in the atmosphere after 10,000 years.[13]

A study from 2013 and endorsed more recently by the Intergovernmental Panel on Climate Change, an international body of the United Nations, found that about 25 percent of the CO_2 emitted into the atmosphere remains after 1,000 years. "I don't want to denigrate forests in the slightest," Danny Cullenward told me, "but a perfect one-to-one exchange with fossil fuels isn't the right way to think about them."[14]

I think the expectations for offsetting with forests are too high. Fires, for example, can risk re-releasing the carbon sequestered in forests. Even if our young trees live their longest lives, those individuals can't keep sequestering carbon every year for the duration that some emitted carbon will linger in the atmosphere. Don't get me wrong; there are still carbon sequestration benefits from trees. But people should be protecting and restoring forests for lots of reasons, and carbon is only one of them.

Cullenward says, and I agree, that too many people entered the scene backward, asking "Where can I find the tonnes?" It's a distraction from the more complex work of protecting, growing, and managing forests well in this rapidly changing world and making these critical land uses viable for local communities. Carbon sequestration will come with the larger package. In the face of the colliding climate and biodiversity crises, we need to become treekeepers, but growing more trees and caring for forests needs to be sustained by many generations to come.

Some companies that are trying to do "better" may be trapped making tallies of their own emissions reductions and offsets elsewhere to reach this one goal of net zero. But if the fossil fuel emissions aren't fully interchangeable with their removal by trees or other means, the end tally in the accounting doesn't equate to what people think it does. It's not *actually* net zero. That reality applies to each one of us too: I cannot walk off the flight from Glasgow, wash the dirt off my own footprint, and say, "Boom! One hundred percent carbon neutral! It's all good."

In 2014, journalist Michael Hobbes wrote an article in the *New Republic* about how big ideas destroy international development, which aims to improve well-being around the world and address the causes and effects of poverty. Titled "Stop Trying to Save the World," it went viral. A dear friend, writer, and expert in environmental justice, Emily Polk, shared it with me. The article had nothing to do with trees or forests, but I read it carefully, seeing countless relevant points from his critical look at why attempting to implement one solution to a problem at scale doesn't work. Hobbes used the example of "PlayPump," a merry-go-round hooked up to water a pump. The big idea was for children to spin around the colorful wheel in villages where water was scarce; the pump would harness their energy to fill nearby tanks and provide communities in sub-Saharan

Africa with fresh water. To make a long story short, the big rollout didn't work. Hobbes called it a "*Mad Libs* version of a narrative we're all familiar with now: Exciting new development idea, huge impact in one location, influx in donor dollars, quick expansion, failure."[15]

I am not suggesting by any means that all reforestation or restoration projects are international development initiatives or that investing a lot of money in the future of forests will lead to failure. But I found his critique of scaling to be relevant. In this "something new under the sun" and even in the "Wild West" of carbon credits, people run a similar risk of looking for one replicable model, one right way. There isn't a prescription; there are only guardrails along many paths forward. Irrespective of size, every endeavor to increase forest cover will be boutique, highly dependent on what was and is possible in place.

I love this line by Hobbes: "What I want to talk shit on is the paradigm of the Big Idea—that once we identify the correct one, we can simply unfurl it on the entire . . . world like a picnic blanket." He writes, "The point is, we don't know what works, where, or why. The only way to find out is to test these models—not just their initial success but afterward, and constantly."

Trees for carbon was and is a big idea. Biodiversity for well-being offers the possibility for revolution. Place by place, plot by plot, we collectively creep forward to a greater reversal, one that remedies the many imbalances between modern society and nature—in addition to the one manifesting in our atmosphere. Reaching great scale may depend more on mainstreaming the process of doing this work "better" and the values motivating it rather than repeatedly implementing the same solution. I sometimes think of this as crafting a mosaic. All the pieces put together create something lasting and beautiful.

"A lot of the best work we can do is going to be diffuse work, not a single project intervention," John-O said. "If you take a potato farm and plant eucalyptus or pine in America, for example, have you made a difference? Someone is still going to grow the potatoes. What's going to happen to those pine trees? So much is out of our control." He admitted to being a bit philosophical in the moment. "But if that shift were part of a movement where every small landowner planted three more trees and had the right understanding about how to care for them, and potatoes could be grown more intensively and replace a high meat diet, which takes up a lot of land use—all that would be a great story. If someone just stops planting potatoes for a little bit and puts some trees in and then twenty years later, they cut the trees and farm again, did we get anything? I don't know. The details really matter and the context matters."

Hobbes wrote in his article, "When you improve something, you change it in ways you couldn't have expected." It reminded me of the concern that solving one problem only creates other issues. Isn't that progress?

I am not willing to forgo the dream for more trees because I know it is a dream for future life—for a better world than what my boy might have without them.

People have often asked me, "Since you're a scientist, what do you think about our climate future?" Now they ask me what will become of all these efforts to restore. During my research, I'd come across images of plantings from a restoration project in Costa Rica. They were captured by satellites and drones. Less than a year and a half after planting, some of the best technology available revealed saplings at one to two meters in height and distinguished their individual leaves. From the bird's-eye view, the saplings looked like plump broccoli crowns. *But are those trees now?* I'd thought. *Is that a forest? Or is that only a hope?*

Success, for me, isn't going to be defined by the number of trees planted or the tonnes of carbon sequestered by forests. Success will be about the great progress achieved toward making better stewardship possible in a world of treekeepers. We have a lot of knowledge and tools available to help. It's dealing with uncertainties and failures and continually reevaluating actions and progress that remains challenging.

Given all the advancements that have occurred in forest monitoring since Virginia Norwood's multispectral scanner system first launched into space in 1972, I know that there will be a very precise record of loss in many places across our planet. But there will also be one of rebirth in others. Every person plays a role in how the extent of degradation and restoration balances out. The inequities of who bears responsibility and how the impacts of "destruction" are distributed still trouble me. So do the imperfect systems of trying to make sustaining nature more economically viable. There isn't a perfect fairy tale ending at a global scale, but I still believe all this work to grow more forests is a pathway to some partial "better."

One morning back home in Montana, while Calder was putting on his sneakers, he told me that he had planted seeds from tiny cones with a friend in the schoolyard. He was on the floor, wrestling with a twisted sock. I leaned down to help. I imagined him digging in the soil during recess, with alder cones in his pockets and loose seed caught in the stitching. I could see the children together, carefully burying their crop amidst the sparce cottonwood and chokecherry trees with both hope and intention.

Where did that seed come from? Where is it going?

I smiled and said to Calder, "Tell me more."

"It didn't work," he replied, more curious than dismayed. I wondered if he'd watered them, if the timing for planting had been suboptimal, or whether they needed more time. I thought about the corotú, moringa, and ceiba seedlings at one of the nurseries on the Azuero Peninsula in Panamá and the recognition that, even with the best care, some of them wouldn't survive.

"I'll keep trying," Calder added, as he secured the last Velcro strap and leapt to his feet.

He didn't know it then, but he was speaking for all the treekeepers and for every one of us trying to figure out how to live in better balance with nature again.

ACKNOWLEDGMENTS

I've never read any of Stephen King's novels, but I loved *On Writing*, his memoir of the craft. One of the ideas he presents from reflections on a life of writing is that every novelist has a single ideal reader—the first trusted reader. That person sneaks into your writing room, then seeps into your head and into your practice; "at various points during the composition of a story, the writer is thinking, 'I wonder what he/she will think when he/she reads this part." I'm not a novelist, but I wrote *Treekeepers* by constantly asking myself, "What will Emily Polk say?" Dear Emily, thank you for being my forever kindred spirit in this writing life, for reading and commenting so thoughtfully on everything, and for always encouraging me to go deeper, share more, and do the critical work of blending multiple perspectives in my reporting.

Years ago, we started a "somewhat secret writing group" at Stanford; then Emily and I wrote about it for *Lit Hub*. Its committed members have kept me afloat through two books now. Poet and nonfiction writer Sara Michas-Martin was a helpful guide through some of my restructuring efforts. It was a treat to exchange chapters with Rob Jackson while he wrote *Into the Clear Blue Sky*, another book that I highly recommend for anyone who wants an in-depth look at the tools we have available to help restore the atmosphere. Richard Nevle somehow found time in his busy teaching and writing schedule to offer extensive comments on nearly every chapter, always identifying something I had overlooked and inspiring me to keep writing forward. I can't wait to read more from the talented writers in this group.

In the summer of 2022, Emily and I led a climate writers' workshop. We created a community of practice by consistently sharing work in the year prior and then convening for a productive and inspirational week of writing in the mountains. Thank you, Christy "Xty" George, Christy Brigham, Lisa Baril, Rob Jackson, and Sally Jackson for joining us. Tatiana Schlossberg, author of *Inconspicuous Consumption*, couldn't attend in person, but I'm grateful for her involvement in our remote sessions and for the conversations we've had about environmental journalism since we first met.

Tom Hayden, Brooke and Terry Tempest Williams, Eric Lambin, Molly Cross, and my former colleagues at the Wildlife Conservation Society, including Nat Moss and Stephen Sautner, were all very encouraging when I was in the brainstorming stage. I wrote a few articles on related topics as I began thinking about a book-length endeavor, and I am grateful to the editors who welcomed those contributions: Kathy Kohm at *Anthropocene Magazine*, Mike Lemonick at *Scientific American*, Seana Quinn at *Emergence Magazine*, and Briana Olson at the *New Farmer's Almanac*. The Department of Earth System Science and the Doerr School of Sustainability at Stanford University have been my intellectual home and community throughout this journey and much more.

Jessica Papin, my thoughtful and talented agent, stood by me and the idea for *Treekeepers* for quite some time—even when the COVID-19 pandemic hit and everything went on hold. I deeply appreciate all your advice over the years. Many thanks to Lara Heimert, TJ Kelleher, Roger Labrie, and Kristen Kim at Basic Books for bringing this story into the world and for making it better through our work together. A book grant from the Alfred P. Sloan Foundation made so much possible. I am eternally grateful to Doron Weber and others at the foundation for supporting my field reporting and enabling me to hire a real deal fact-checker. The real deal is Sean Lavery, the former head research editor at the *New Yorker*. I am so grateful that you said yes after reviewing my early draft and then tackled the mission with gusto. It was such a pleasure to work with you and learn from you, Sean. You are the best of the best. Tatiana, thanks for introducing us.

Catherine Kuhn rallied to be another early reader; her insightful feedback helped resolve some lingering issues and fill important gaps. Lorenzo Rosenzweig Pasquel approached me after I gave a keynote at a conservation symposium in Denver. We were talking about conservation finance and reforestation in Mexico, but he mentioned his side passion for illustration. Lorenzo offered to draw anything I needed for this book and promptly started sketching. Tanvi Dutta Gupta helped finalize the endnotes. After Lisa Baril finished writing *The Age of Melt*, a stunning account of ice patch archeology and the stories revealed by retreating glaciers, she volunteered to help me too. It was a sprint, given my editing timeline and the impending birth of our second child, Arden.

Writing becomes such an all-encompassing experience for me. If I'm not at my desk, the other places where I write sustain the steady rhythm. I appreciate the little cafés that brought respite from the solitude and the kind friends and contacts who welcomed me into their sweet spaces to work. A few highlights: Libby's cozy place in Bear Canyon; the main street office for the publishing team of *Big Sky Journal*; Greg's yellow house beside Beale Park with the chickens roaming outside; our trailside rental home on Cherry Street; one small cabin in Roberts, Montana; Treeline Coffee in Bozeman (how could I *not* write there?); Kaya's Café and Usha's Yurt in Hawai'i; and Matt's parents' thatched-roof home in England. Thank you to Marm Kilpatrick's family as well as Noa Lincoln and Dana Shapiro for making my trip to Hawai'i possible on a shoestring budget. Becky Taylor kindly hosted me in Seattle.

Laura Rocchio at the National Aeronautics and Space Administration was helpful with fact-checking the Landsat history. Thank you, Sacha Spector, for your insights on nature-based solutions and policy over our years of working together. Alvaro Umaña became an unexpected mentor and colleague after I interviewed him a few times. He thoughtfully reviewed numerous sections and contributed to my research on the earliest efforts to pay for ecosystem services and offset emissions. Grayson Bagley provided an additional check on content related to credits. John Weir had many leads for my reporting in the United Kingdom. In trust, Paul Faeth sent me his personal 35mm slides from the AES project in Guatemala. I had intended to use them for description. It still breaks my heart that UPS lost them, and I wonder if maybe, just maybe, they'll show up at my door one day. My mother gets an award for her constant encouragement and for spending two days in a storage unit to find my old photographs of the cross in the tree.

I wish I could have included everyone I interviewed and what they shared with me too. I apologize to those who didn't make it into this book by name or direct reference to their work. Please know that you are still part of this story and, more importantly, part of this race for "better."

Matt, it must have been a strange experience living with an obsessed researcher and writer as your partner who also wouldn't let you read anything until the book was nearly finished. Thank you for believing in me and supporting the process, even when it took me away from our family. Our regular conversations helped unstick me more times than you know, and the story improved because you are a better editor than I knew.

I will never forget the first time Calder asked how many words I had written on one random day. He was four years old, and the inquiry then became a ritual. I was discouraged that afternoon and answered, "Barely 400." He exclaimed, "What? 400?! You wrote 400 *whole* words? That's amazing." I think he was picturing me slowly scrawling letters with crayons all day. Calder, how special it was to read sections of *Treekeepers* with you at your young age. Your drawings of Carboniferous forests brightened my office. When you are older and you read this book, I hope you see how grateful I am for your perspective and the many ways you show me beauty.

INTERVIEWS

A woman once described a friend of hers as being such a keen listener that even the trees leaned toward her, as if they were speaking their innermost secrets into her listening ears. Over the years, I've envisioned that woman's silence, a hearing full and open enough that the world told her stories. The green leaves turned toward her, whispering tales of soft breezes and the murmurs of leaf against leaf.

—Linda Hogan, from "A Different Yield"

In gratitude to Tom Ager, Sally Aitken, Sehr Ali, Leander Anderegg, Christa Anderson, Melissa Annetts, Yustina Artati, Caroline Ayre, Andy Baker, Alaine Ball, Himlal Baral, Jean-François Bastin, Craig Beatty, Ross Bernet, Christophe Besacier, Johann Besserer, Ursula Biemann, Aron Boettcher, Jean Philippe J. Bouancheau, Kevin Boyce, Pedro Brancalion, Greg Bratman, Elinor Breman, Christy Brigham, Kathleen Buckingham, Mike Carey, Troy Carter, Marian Chau, Robin Chazdon, Victor Clements, Andrew Coates, Susan Cook-Patton, Meredith Cornett, Nigel Couch, Tom Crowther, Danny Cullenward, Jad Daley, Luc Daniel, Sharon Danks, Lindsay Darling, Chris David, Daisy Dent, Bill DiMichele, Richard Donovan, Lee Dukatkin, Heidi Dungey, Matt Elliot, Ailene Ettinger, Paul Faeth, Jamie Fairbairn, Jacques Fillâtre, Sheilagh Fitzpatrick, Julian Fox, Hector Frías, Jason Funk, Hilary Galloway, Dan Gaudet, Malcolm Porteus González, Hedley Grantham, Stephanie Greene, Mary Hammes,

James Hand, Matt Hansen, Kate A. Hardwick, Eleanor Harris, Becky Hart, Xavier Hatchondo, Andrew Heald, Molly Henry, Mariana Lopez de Heredia, Matt Hill, Olivia Hill, Dryw Jones, Marie-Nöelle Keijzer, Brian Kittler, David Knott, Eric Lambin, Bryn Lawrence, Yee Lee, Patrick Leung, Jamie Lewis, Noa Lincoln, Damon Little, Roberto Machazek, Will MacKenzie, Jeremy Manion, Mark McPherson, Patrick Meyfroidt, Jake Miesbauer, Isabel Montañez, Camilo Mora, Greg Nagle, Ezra Neale, Coby Needle, Matt Nelsen, Cara Nelson, Jonny Nesmith, Sarah Nicholas, John-O Niles, Naomi Norwood, Virginia Norwood, Kevin O'Hara, Greg O'Neill, Mike Osborne, Andrea Paz, Alexandra Leeds Perkins, Trinity Pierce, Suzanne Prober, John Purslow, Alan Rasmussen, Tim Rayden, Katie Reytar, Christy Rollinson, Clara Rowe, Tom Rudel, Jeremy Caves Rugenstein, Minister René Castro Salazar, Allyson Salisbury, Miriam Samaniego, Roger Sant, Dmitry Schepaschenko, Steve Scott, Katy Shallows, Gancho Slavov, Paul Smith, Xiaopeng Song, Sheryl Sturges, Sam Suarez, Hubert Szczygieł, Pooja Tandon, Keali'i Thoene, Lehua Todero, Randall Tolpinrud, John Townshend, Alvaro Umaña, Marcos Valderrábano, Victor Vankus, Sandra Vásquez de Zambrano, Jill Wagner, Bethanie Walder, Sarah Walker, Chris Waterfield, Will Weaver, John Weir, Leland Werden, Tom Whitman, Brooke Williams, Rivah Winter, Alex Woods, Lowell Wyse, César Zambrano

EPIGRAPH CREDITS

Mary Oliver, excerpt from "When I Am Among the Trees." Reprinted with the permission of the Charlotte Sheedy Literary Agency as agent for the author. Copyright © 2006, 2010, 2017 by Mary Oliver with the permission of Bill Reichblum.

Philip Larkin, excerpt from "The Trees," from *The Complete Poems of Philip Larkin* by Philip Larkin, edited by Archie Burnett. Copyright © 2012 by The Estate of Philip Larkin. Reprinted with permission of Farrar, Straus, and Giroux. All rights reserved.

Ada Limón, excerpt from "Instructions on Not Giving Up," from *The Carrying*. Copyright © 2018 by Ada Limón. Reprinted with the permission of the Permissions Company LLC on behalf of Milkweed Editions, milkweed.org.

Mark Strand, "Keeping Things Whole," from *Selected Works of Mark Strand* by Mark Strand. Copyright © 1979, 1980 by Mark Strand. Used with the permission of Alfred A. Knopf, an imprint of the Knopf Doubleday Publishing Group, a division of Penguin House, LLC. All rights reserved.

Linda Hogan, "A Different Yield," from *Selected Works of Mark Strand* by Linda Hogan. Copyright © 1995 by Linda Hogan. Reprinted with the permission of the author.

NOTES

Prologue

1. the forests that remain today . . . Harris, Nancy L., David A. Gibbs, Alessandro Baccini, Richard A. Birdsey, Sytze de Bruin, Mary Farina, Lola Fatoyinbo, et al. "Global Maps of Twenty-First Century Forest Carbon Fluxes." *Nature Climate Change* 11, no. 3 (March 2021): 234–40. https://doi.org/10.1038/s41558-020-00976-6; Heinrich, Viola H. A., Ricardo Dalagnol, Henrique L. G. Cassol, Thais M. Rosan, Catherine Torres de Almeida, Celso H. L. Silva Junior, Wesley A. Campanharo, et al. "Large Carbon Sink Potential of Secondary Forests in the Brazilian Amazon to Mitigate Climate Change." *Nature Communications* 12, no. 1 (March 19, 2021): 1785. https://doi.org/10.1038/s41467-021-22050-1; Pan, Yude, Richard A. Birdsey, Jingyun Fang, Richard Houghton, Pekka E. Kauppi, Werner A. Kurz, Oliver L. Phillips, et al. "A Large and Persistent Carbon Sink in the World's Forests." *Science* 333, no. 6045 (August 19, 2011): 988–93. https://doi.org/10.1126/science.1201609.

Chapter 1: Baseline

1. Timberland launched a campaign . . . Aghadjanian, Nina. "Timberland Pledges to Plant 50 Million Trees in Latest Campaign Merging Marketing and Sustainability." *AList*, September 9, 2019; "Timberland Commits to Plant 50 Million Trees over Next Five Years." *Businesswire*, September 5, 2019. www.businesswire.com/news/home/20190905005555/en/Timberland-Commits-to-Plant-50-Million-Trees-Over-Next-Five-Years.

2. When she joined the antenna lab . . . Dragoon, Alice. "The Woman Who Brought Us the World." *MIT News*, June 29, 2021. www.technologyreview.com/2021/06/29/1025732/the-woman-who-brought-us-the-world.

3. Virginia's scanner could capture . . . "Landsat Data." USGS Fact Sheet. US Geological Survey, December 1997. https://pubs.usgs.gov/fs/0084-97/report.pdf.

4. *New York Times* proclaimed . . . Rensberger, Boyce. "An Earth-Exploring Satellite Is Orbited." *New York Times*, July 24, 1972.

5. Within days of the launch . . . Landsat Missions. "Landsat 1." US Geological Survey. Accessed December 18, 2023. www.usgs.gov/landsat-missions/landsat-1.

6. The Rondônia images revealed . . . Goward, Samuel N., Darrel L. Williams, Terry Arvidson, Lauren E. P. Rocchio, James R. Irons, Carol A. Russell, and Shaida S. Johnston. "Landsat's Enduring Legacy." American Society for Photogrammetry and Remote Sensing, 2017. www.asprs.org/landsat.

7. one million hours of computing . . . Hansen, M. C., P. V. Potapov, R. Moore, M. Hancher, S. A. Turubanova, A. Tyukavina, D. Thau, et al. "High-Resolution Global Maps of 21st-Century Forest Cover Change." *Science* 342, no. 6160 (November 15, 2013): 850–53. https://doi.org/10.1126/science.1244693.

8. deforestation in Brazil had slowed . . . Nepstad, Daniel, Britaldo S. Soares-Filho, Frank Merry, André Lima, Paulo Moutinho, John Carter, Maria Bowman, et al. "The End of Deforestation in the Brazilian Amazon." *Science* 326, no. 5958 (December 4, 2009): 1350–51. https://doi.org/10.1126/science.1182108.

9. In 2018, scientists came forth . . . Song, Xiao-Peng, Matthew C. Hansen, Stephen V. Stehman, Peter V. Potapov, Alexandra Tyukavina, Eric F. Vermote, and John R. Townshend. "Global Land Change

from 1982 to 2016." *Nature* 560, no. 7720 (August 2018): 639–43. https://doi.org/10.1038/s41586-018-0411-9.

10. the most recent report . . . *Global Forest Resources Assessment 2020*. FAO, 2020. https://doi.org/10.4060/ca8753en.

11. result of three trillion trees . . . Crowther, T. W., H. B. Glick, K. R. Covey, C. Bettigole, D. S. Maynard, S. M. Thomas, J. R. Smith, et al. "Mapping Tree Density at a Global Scale." *Nature* 525, no. 7568 (September 2015): 201–5. https://doi.org/10.1038/nature14967; made headlines, of course . . . Jacobs, Suzanne. "Earth Has Way More Trees Than We Thought, but Not Nearly as Many as It Used To." *Grist*, September 3, 2015. https://grist.org/science/earth-has-way-more-trees-than-we-thought-but-not-nearly-as-many-as-it-used-to.

Chapter 2: The Potential

1. this is a beautiful thing . . . Nikiforuk, Andrew. "If We Plant Billions of Trees to Save Us, They Must Be Native Trees." *The Tyee*, February 28, 2020. https://thetyee.ca/News/2020/02/28/If-We-Plant-Trees-They-Must-Be-Native-Trees; Lang, Chris. "REDD-Monitor." Substack. Accessed December 18, 2023. https://reddmonitor.substack.com/embed.

2. the video on YouTube . . . "The Global Tree Restoration Potential." Video posted to YouTube by Crowther Lab, October 8, 2020. www.youtube.com/watch?v=v30tP-lrI-w.

3. the drylands forest study . . . Bastin, Jean-François, Nora Berrahmouni, Alan Grainger, Danae Maniatis, Danilo Mollicone, Rebecca Moore, Chiara Patriarca, et al. "The Extent of Forest in Dryland Biomes." *Science* 356, no. 6338 (May 12, 2017): 635–38. https://doi.org/10.1126/science.aam6527.

4. published in *Science* . . . Bastin, Jean-Francois, Yelena Finegold, Claude Garcia, Danilo Mollicone, Marcelo Rezende, Devin Routh, Constantin M. Zohner, et al. "The Global Tree Restoration Potential." *Science* 365, no. 6448 (July 5, 2019): 76–79. https://doi.org/10.1126/science.aax0848.

5. clarified their calculation . . . "Erratum for the Report: 'The Global Tree Restoration Potential' by Bastin, J.-F., Y. Finegold, C. Garcia, D. Mollicone, M. Rezende, D. Routh, C. M. Zohner, T. W. Crowther and for the Technical Response 'Response to Comments on "The Global Tree Restoration Potential"' by Bastin, J.-F., Y. Finegold, C. Garcia, N. Gellie, A. Lowe, D. Mollicone, M. Rezende, D. Routh, M. Sacande, B. Sparrow, C. M. Zohner, T. W. Crowther." *Science* 368, no. 6494 (May 29, 2020): eabc8905. https://doi.org/10.1126/science.abc8905; the airborne fraction, 45 percent . . . Keeling, Charles D., T. P. Whorf, M. Wahlen, and J. van der Plichtt. "Interannual Extremes in the Rate of Rise of Atmospheric Carbon Dioxide Since 1980." *Nature* 375, no. 6533 (1995): 666–70.

6. The results were twisted . . . "Could 1 Trillion Trees Stop Climate Change?" *DW*, July 4, 2019. www.dw.com/en/planting-1-trillion-trees-could-stop-climate-change-argues-study/a-49478494; Lewis, Sophie. "Planting a Trillion Trees Could Be the 'Most Effective Solution' to Climate Change, Study Says." *CBS News*, July 8, 2019. www.cbsnews.com/news/planting-a-trillion-trees-could-be-the-most-effective-solution-to-climate-change; Carrington, Damian. "Tree Planting 'Has Mind-Blowing Potential' to Tackle Climate Crisis." *The Guardian*, July 4, 2019, sec. Environment. www.theguardian.com/environment/2019/jul/04/planting-billions-trees-best-tackle-climate-crisis-scientists-canopy-emissions.

7. A group of scientists commenting . . . Ellis, Erle C., Mark Maslin, and Simon Lewis. "Planting Trees Won't Save the World." *New York Times*, February 12, 2020, sec. Opinion. www.nytimes.com/2020/02/12/opinion/trump-climate-change-trees.html.

8. *Rolling Stone* called the idea . . . Goodell, Jeff. "Why Planting Trees Won't Save Us." *Rolling Stone*, June 25, 2020. www.rollingstone.com/politics/politics-features/tree-planting-wont-stop-climate-crisis-1020500.

9. Some critics noted . . . Luedeling, Eike, Jan Börner, Wulf Amelung, Katja Schiffers, Keith Shepherd, and Todd Rosenstock. "Forest Restoration: Overlooked Constraints." *Science* 366, no. 6463 (October 18, 2019): 315. https://doi.org/10.1126/science.aay7988.

10. Other experts were concerned . . . Veldman, Joseph W., Julie C. Aleman, Swanni T. Alvarado, T. Michael Anderson, Sally Archibald, William J. Bond, Thomas W. Boutton, et al. "Comment on 'The Global Tree Restoration Potential.'" *Science* 366, no. 6463 (October 18, 2019): eaay7976. https://doi.org/10.1126/science.aay7976.

11. Ecological restoration is more than . . . "What Is Ecological Restoration?" Society for Ecological Restoration. Accessed December 18, 2023. https://ser-rrc.org/what-is-ecological-restoration.

12. A study, coauthored by . . . Walker, Wayne S., Seth R. Gorelik, Susan C. Cook-Patton, Alessandro Baccini, Mary K. Farina, Kylen K. Solvik, Peter W. Ellis, et al. "The Global Potential for Increased Storage of Carbon on Land." *Proceedings of the National Academy of Sciences* 119, no. 23 (June 7, 2022): e2111312119. https://doi.org/10.1073/pnas.2111312119.

13. forests in the United States remove . . . Domke, Grant M., Sonja N. Oswalt, Brian F. Walters, and Randall S. Morin. "Tree Planting Has the Potential to Increase Carbon Sequestration Capacity of Forests in the United States." *Proceedings of the National Academy of Sciences* 117, no. 40 (October 6, 2020): 24649–51. https://doi.org/10.1073/pnas.2010840117; Durkay, Jocelyn, and Jennifer Schultz. "The Role of Forests in Carbon Sequestration and Storage." National Conference of State Legislatures, January 1, 2016. www.ncsl.org/environment-and-natural-resources/the-role-of-forests-in-carbon-sequestration-and-storage.

14. area about twice the size of Oregon . . . Cook-Patton, Susan C., Trisha Gopalakrishna, Adam Daigneault, Sara M. Leavitt, Jim Platt, Sabrina M. Scull, Oyut Amarjargal, et al. "Lower Cost and More Feasible Options to Restore Forest Cover in the Contiguous United States for Climate Mitigation." *One Earth* 3, no. 6 (2020): 739–52. https://doi.org/10.1016/j.oneear.2020.11.013.

15. the Global Carbon Project released . . . Friedlingstein, Pierre, Michael O'Sullivan, Matthew W. Jones, Robbie M. Andrew, Dorothee C. E. Bakker, Judith Hauck, Peter Landschützer, et al. "Global Carbon Budget 2023." *Earth System Science Data* 15, no. 12 (December 5, 2023): 5301–69. https://doi.org/10.5194/essd-15-5301-2023. Note the studies used by the Global Carbon Project for its 2023 update: IPCC. *Climate Change 2023: Synthesis Report. Contribution of Working Groups I, II and III to the Sixth Assessment Report of the Intergovernmental Panel on Climate Change*, ed. Core Writing Team, H. Lee, and J. Romero (Geneva: IPCC, 2023), 35–115. doi: 10.59327/IPCC/AR6-9789291691647; Lamboll, Robin D., Zebedee R. J. Nicholls, Christopher J. Smith, Jarmo S. Kikstra, Edward Byers, and Joeri Rogelj. "Assessing the Size and Uncertainty of Remaining Carbon Budgets." *Nature Climate Change* (2023): 1–8; Forster, Piers M., Christopher J. Smith, Tristram Walsh, William F. Lamb, Robin Lamboll, Mathias Hauser, Aurélien Ribes, Debbie Rosen, et al. "Indicators of Global Climate Change 2022: Annual Update of Large-Scale Indicators of the State of the Climate System and Human Influence." *Earth System Science Data* 15, no. 6 (2023): 2295–327.

16. Corporate interest intensified . . . Au-Yeung, Angel. "Jeff Bezos Announces Nearly $800 Million in Donations to Climate Change Organizations." *Forbes*. Accessed December 18, 2023. www.forbes.com/sites/angelauyeung/2020/11/16/jeff-bezos-announces-nearly-800million-in-donations-to-climate-change-organizations; Welch, Craig. "Tree-Planting Projects Abound. Which Should You Support?" *National Geographic*, May 16, 2022. www.nationalgeographic.com/environment/article/tree-planting-projects-abound-which-should-you-support; Holl, Karen D., and Pedro H. S. Brancalion. "Which of the Plethora of Tree-Growing Projects to Support?" *One Earth* 5, no. 5 (May 20, 2022): 452–55. https://doi.org/10.1016/j.oneear.2022.04.001; Amazon Staff. "Alexa, Grow a Tree." Amazon, April 11, 2022. www.aboutamazon.com/news/devices/alexa-grow-a-tree.

17. created some strange bedfellows . . . Darby, Dabreon. "National Public Lands Day 2020." DENIX, April 31, 2021. www.denix.osd.mil/legacy/nr-legacy-project-deliverables/fy2020-fy2021/deliverable/legacy-neef-npld-20-department-of-defense-national-public-lands-day-2020-final-report/NEEF-NPLD-20_Final%20Report_FINAL.pdf; his Build Back Better . . . Chamberlain, Samuel. "Biden, Dems' $3.5T Bill Includes Money for 'Tree Equity,' Bias Training." *New York Post*, September 27, 2021. https://nypost.com/2021/09/27/biden-dems-3-5t-bill-includes-money-for-tree-equity-bias-training; Jones, Benji. "Biden's $27 BillionBet on Forests." *Vox*, October 29, 2021. www.vox.com/down-to-earth/2021/10/29/22752019/build-back-better-bill-climate-forest-fires-wildlife; Reuters. "Chevron Would Rather Pay Dividends Than Invest in Wind and Solar—CEO." *Reuters*, September 16, 2021. www.reuters.co/business/energy/chevron-ceo-says-dividend-shareholders-better-than-investing-wind-solar-2021-09-15.

18. Tom Crowther had also become . . . "UN Decade on Restoration." www.decadeonrestoration.org; "Steve Jobs of Ecology". . . Greenfield, Patrick. "'I've Never Said We Should Plant a Trillion Trees': What Ecopreneur Thomas Crowther Did Next." *The Guardian*, September 1, 2021, sec. Environment. www.theguardian.com/environment/2021/sep/01/ive-never-said-we-should-plant-a-trillion-trees-what-ecopreneur-thomas-crowther-did-next-aoe.

19. The media had turned . . . Irwin, Aisling. "The Ecologist Who Wants to Map Everything." *Nature* 573, no. 7775 (September 20, 2019): 478–81. https://doi.org/10.1038/d41586-019-02846-4; "Catchy

findings have propelled" . . . Popkin, Gabriel. "Catchy Findings Have Propelled This Young Ecologist to Fame—and Enraged His Critics," *Science*, October 24, 2019. https://www-science-org.stanford.idm.oclc.org/content/article/catchy-findings-have-propelled-young-ecologist-fame-and-enraged-his-critics 2/10.

20. "Look around you" . . . Weisman, Alan. *The World Without Us* (New York: Thomas Dunne Books, 2007), 4.

Chapter 3: From Long Ago

1. the Carboniferous, which began . . . MacDonald, James. "The Age of Amphibians." *JSTOR Daily*, December 10, 2015. https://daily.jstor.org/the-age-of-amphibians; "Carboniferous Period Information and Prehistoric Facts." *National Geographic*. Accessed December 18, 2023. www.nationalgeographic.com/science/article/carboniferous; Harrison, Jon F., Alexander Kaiser, and John M. VandenBrooks. "Atmospheric Oxygen Level and the Evolution of Insect Body Size." *Proceedings of the Royal Society B: Biological Sciences*. Accessed December 18, 2023. https://doi.org/10.1098/rspb.2010.0001; vegetative communities changed in waves . . . Montañez, Isabel Patricia. "A Late Paleozoic Climate Window of Opportunity." *Proceedings of the National Academy of Sciences* 113, no. 9 (March 2016): 2334–36. https://doi.org/10.1073/pnas.1600236113.

2. lycopsids of the Coal Age . . . Boyce, C. Kevin, and William A. DiMichele. "Arborescent Lycopsid Productivity and Lifespan: Constraining the Possibilities." *Review of Palaeobotany and Palynology* 277 (April 2016): 97–110. doi: https://doi.org/10.1016/j.revpalbo.2015.10.007; Those extensive root structures . . . DiMichele, William A., Richard M. Bateman, Gar W. Rothwell, Ivo A. P. Duijnstee, Scott D. Elrick, and Cynthia V. Looy. "Stigmaria: A Review of the Anatomy, Development, and Functional Morphology of the Rootstock of the Arboreous Lycopsids." *International Journal of Plant Sciences* 183, no. 6 (July 2022): 493–534. https://doi.org/10.1086/720641.

3. Between about 330 and 260 million years . . . Montañez, Isabel Patricia. "A Late Paleozoic Climate Window of Opportunity." *Proceedings of the National Academy of Sciences* 113, no. 9 (March 2016): 2334–36. https://doi.org/10.1073/pnas.1600236113; For about nine million years . . . Boyce, C. Kevin, Mike Abrecht, Dong Zhou, and P. U. P. A. Gilbert. "X-Ray Photoelectron Emission Spectromicroscopic Analysis of Arborescent Lycopsid Cell Wall Composition and Carboniferous Coal Ball Preservation." *International Journal of Coal Geology* [Hermann W. Pfefferkorn Commemorative Volume] 83, no. 2 (August 1, 2010): 146–53. https://doi.org/10.1016/j.coal.2009.10.008.

4. factors that enabled tropical organic matter . . . Nelsen, Matthew P., William A. DiMichele, Shanan E. Peters, and C. Kevin Boyce. "Delayed Fungal Evolution Did Not Cause the Paleozoic Peak in Coal Production." *Proceedings of the National Academy of Sciences* 113, no. 9 (March 2016): 2442–47. https://doi.org/10.1073/pnas.1517943113.

5. dirtiest of the fossil fuels . . . Edwards, Gareth A. S. "Coal and Climate Change." *WIREs Climate Change* 10, no. 5 (July 17, 2019). https://wires.onlinelibrary.wiley.com/doi/abs/10.1002/wcc.607; Gohlke, Julia M., Reuben Thomas, Alistair Woodward, Diarmid Campbell-Lendrum, Annette Prüss-Üstün, Simon Hales, and Christopher J. Portier. "Estimating the Global Public Health Implications of Electricity and Coal Consumption." *Environmental Health Perspectives* 119, no. 6 (June 2011): 821–26. https://doi.org/10.1289/ehp.119-821; Between 1850 and 2020 . . . Friedlingstein, Pierre, Matthew W. Jones, Michael O'Sullivan, Robbie M. Andrew, Dorothee C. E. Bakker, Judith Hauck, Corinne Le Quéré, et al. "Global Carbon Budget 2021." *Earth System Science Data* 14, no. 4 (April 26, 2022): 1917–2005. https://doi.org/10.5194/essd-14-1917-2022.

Chapter 4: Something New Under the Sun

1. the term *forest transition* . . . Rudel, Thomas K., Laura Schneider, and Maria Uriarte. "Forest Transitions: An Introduction." *Land Use Policy* 27, no. 2 (April 1, 2010): 95–97. https://doi.org/10.1016/j.landusepol.2009.09.021.

2. studying the causes of forest decline . . . Geist, Helmut J., and Eric F. Lambin. "Proximate Causes and Underlying Driving Forces of Tropical Deforestation." *BioScience* 52, no. 2 (2002): 143. https://doi.org/10.1641/0006-3568(2002)052[0143:PCAUDF]2.0.CO;2.

3. Mather had died . . . "The Obituary Notice of Alexander Sandy Smith MATHER." Funeral Notices. Accessed December 19, 2023. https://funeral-notices.co.uk/notice/MATHER/3639291.

4. Scotland's forest transition . . . "History of Scotland's Woodlands." NatureScot, January 27, 2023. www.nature.scot/professional-advice/land-and-sea-management/managing-land/forests-and-woodlands/history-scotlands-woodlands; But by around 1600 . . . Smout, T. Christopher, and Scottish Natural Heritage (Agency). *Scotland Since Prehistory: Natural Change and Human Impact* (Newbattle: Scottish Cultural Press, 1993).

5. "There is no doubt . . ." Mather, A. S. "Forest Transition Theory and the Reforesting of Scotland." *Scottish Geographical Journal* 120, no. 1–2 (January 1, 2004): 83–98. https://doi.org/10.1080/00369220418737194.

6. In the Netherlands, forest area grew . . . Grandjean, A. J. "Forets domaniales et politique forestiere neerlandaise depuis 1945." *Revue forestiere francaise* 39 (1987): 219–30; Forests occupied nearly 20 percent . . . du Saussay, Christian. *Land Tenure Systems and Forest Policy*. FAO Legislative Study (Rome: Food and Agriculture Organization of the United Nations, 1987); In Denmark, forest area had . . . Knudsen, F. D. "Forestry: Denmark." In *European Environmental Yearbook*, 251–52 (London: DocTer, 1987); Totman, Conrad. "From Exploitation to Plantation Forestry in Early Modern Japan." In *History of Sustained Yield Forestry: A Symposium*, ed. Harold K. Steen (Santa Cruz, CA: Forest History Society, 1984): 370–80; Zhang, Yaoqi. "Deforestation and Forest Transition: Theory and Evidence in China." In *World Forests from Deforestation to Transition?*, ed. Matti Palo and Heidi Vanhanen, 2:41–65 (Dordrecht: Springer Netherlands, 2000). https://doi.org/10.1007/978-94-010-0942-3_3; Pfaff, Alexander, and Robert Walker. "Regional Interdependence and Forest 'Transitions': Substitute Deforestation Limits the Relevance of Local Reversals." *Land Use Policy* 27, no. 2 (April 1, 2010): 119–29. https://doi.org/10.1016/j.landusepol.2009.07.010; workers from the Civilian Conservation Corps . . . Reed, Leslie. "FDR's 'Great Wall of Trees' Continues to Provide Lessons." *Nebraska Today*, May 18, 2017. https://news.unl.edu/newsrooms/today/article/fdr-s-great-wall-of-trees-continues-to-provide-lessons.

7. the loss of Scotland's forests . . . Mather, A. S. "Forest Transition Theory and the Reforesting of Scotland." *Scottish Geographical Journal* 120, no. 1–2 (January 1, 2004): 83–98. https://doi.org/10.1080/00369220418737194; "Forestry Act of 1919." *London Gazette*, March 26, 1920.

8. "Little attention was devoted" . . . Mather, A. S. "Forest Transition Theory and the Reforesting of Scotland." *Scottish Geographical Journal* 120, no. 1–2 (January 1, 2004): 83–98. https://doi.org/10.1080/00369220418737194.

9. first global wall-to-wall map . . . Lesiv, Myroslava, Dmitry Schepaschenko, Marcel Buchhorn, Linda See, Martina Dürauer, Ivelina Georgieva, Martin Jung, et al. "Global Forest Management Data for 2015 at a 100 m Resolution." *Scientific Data* 9, no. 1 (May 10, 2022): 199. https://doi.org/10.1038/s41597-022-01332-3.

10. fears of dependency . . . Rudel, Thomas K. "Shocks, States, and Societal Corporatism: A Shorter Path to Sustainability?" *Journal of Environmental Studies and Sciences* 9, no. 4 (December 1, 2019): 429–36. https://doi.org/10.1007/s13412-019-00560-1.

11. In one of the many articles . . . Mather, Alexander. "The Forest Transition." *Area* (1992): 367–79.

12. echoed back to other writings . . . Friedrich, Ernst. "Wesen und Geographische Verbreitung der 'Raubwirtschaft.'" *Petermanns Mitteilungen* 50 (1904): 68–79, 92–95.

13. Xiaopeng Song and his colleagues . . . Song, Xiao-Peng, Matthew C. Hansen, Stephen V. Stehman, Peter V. Potapov, Alexandra Tyukavina, Eric F. Vermote, and John R. Townshend. "Global Land Change from 1982 to 2016." *Nature* 560, no. 7720 (August 2018): 639–43. https://doi.org/10.1038/s41586-018-0411-9.

14. drop in global poverty estimates . . . "Lifting 800 Million People Out of Poverty—New Report Looks at Lessons from China's Experience." World Bank, April 1, 2022. www.worldbank.org/en/news/press-release/2022/04/01/lifting-800-million-people-out-of-poverty-new-report-looks-at-lessons-from-china-s-experience; rates of poverty in sub-Saharan Africa . . . Aikins, Enoch Randy, and Jacobus du Toit McLachlan. "Africa Is Losing the Battle Against Extreme Poverty." ISS Africa, July 13, 2022. https://issafrica.org/iss-today/africa-is-losing-the-battle-against-extreme-poverty.

15. National policies, for example . . . Meyfroidt, Patrick, and Eric F. Lambin. "Global Forest Transition: Prospects for an End to Deforestation." *Annual Review of Environment and Resources* 36, no. 1 (November 21, 2011): 343–71. https://doi.org/10.1146/annurev-environ-090710-143732.

16. largest-known living single-stem tree . . . "Largest Living Tree (Volume)." Guinness World Records. Accessed December 19, 2023. www.guinnessworldrecords.com/world-records/largest-living-tree-; "The

General Sherman Tree." National Park Service, September 27, 2023. www.nps.gov/seki/learn/nature/sherman.html; "General Sherman (Tree)." Wikipedia, December 17, 2023. https://en.wikipedia.org/w/index.php?title=General_Sherman_(tree)&oldid=1190411377.

Chapter 5: A Forest Is

1. 800 definitions of forests . . . Intergovernmental Panel on Climate Change, *Expert Meeting on Harmonizing Forest-Related Definitions for Use by Various Stakeholders* (Rome: UNFAO, 2002); range of estimates of global forest area . . . Sexton, Joseph O., Praveen Noojipady, Xiao-Peng Song, Min Feng, Dan-Xia Song, Do-Hyung Kim, Anupam Anand, et al. "Conservation Policy and the Measurement of Forests." *Nature Climate Change* 6, no. 2 (February 2016): 192–96. https://doi.org/10.1038/nclimate2816.

2. A forest is a biological community . . . Shvidenko, Anatoly, C. V. Barber, and R. Persson. "Forests and Woodland Systems." In *Ecosystems and Human Well-Being: Current State and Trends*. Millennium Ecosystem Assessment 1 (Washington, DC: Island Press, 2005): 587–621.

3. The Food and Agriculture Organization . . . Chazdon, Robin L., Pedro H. S. Brancalion, Lars Laestadius, Aoife Bennett-Curry, Kathleen Buckingham, Chetan Kumar, Julian Moll-Rocek, et al. "When Is a Forest a Forest? Forest Concepts and Definitions in the Era of Forest and Landscape Restoration." *Ambio* 45, no. 5 (2016): 538–50.

4. a photo of verdant valley . . . Lund, H. G. "What Is a Forest? Definitions Do Make a Difference: An Example from Turkey." *Avrasya Terim Dergisi* 2, no. 1 (2014): 1–8.

5. Norwood's sensors on Landsat 1 . . . "What Are the Band Designations for the Landsat Satellites?" US Geological Survey. www.usgs.gov/faqs/what-are-band-designations-landsat-satellites; Earth Resources Observation and Science (EROS) Center. "USGS EROS Archive—Landsat Archives—Landsat 1-5 Multispectral Scanner (MSS) Level-1 Data Products." Tiff. US Geological Survey, 2018. https://doi.org/10.5066/F7H994GQ.

6. "Names," she writes . . . Kimmerer, Robin. *Braiding Sweetgrass* (Minneapolis: Milkweed Editions, 2013), 56.

7. An intact forest landscape . . . "Intact Forest Landscapes," https://intactforests.org.

8. The list of services . . . Watson, J. E., T. Evans, O. Venter, B. Williams, A. Tulloch, C. Stewart, and D. Lindenmayer. "The Exceptional Value of Intact Forest Ecosystems." *Nature Ecology and Evolution* 2, no. 4 (2018): 599–610.

9. a pixel inside of each of those forests . . . Grantham, H. S., A. Duncan, T. D. Evans, K. R. Jones, H. L. Beyer, R. Schuster, and J. E. M. Watson. "Anthropogenic Modification of Forests Means Only 40% of Remaining Forests Have High Ecosystem Integrity." *Nature Communications* 11, no. 1 (2020): 1–10.

Chapter 6: Choosing the Trees

1. British Columbia has been well positioned . . . "FACTSHEET: Reforestation in B.C." BC Gov News, April 7, 2017. https://news.gov.bc.ca/factsheets/factsheet-reforestation-in-bc.

2. eight billion trees had been planted . . . "Forests for Tomorrow." Province of British Columbia. Accessed December 19, 2023. https://www2.gov.bc.ca/assets/gov/environment/natural-resource-stewardship/land-based-investment/forests-for-tomorrow/fft_booklet.pdf.

3. A 2021 documentary reported . . . *Forest for the Trees: The Tree Planters*. Green Planet Films, 2021. www.forestforthetreesdocumentary.com.

4. A low oxygen environment . . . Graziano, Jack, and Macarena Farcuh. "Controlled Atmosphere Storage of Apples." University of Maryland Extension, September 10, 2021. https://extension.umd.edu/resource/controlled-atmosphere-storage-apples.

5. studies in Portugal and California . . . da Silva, Luís P., Ruben H. Heleno, José M. Costa, Mariana Valente, Vanessa A. Mata, Susana C. Gonçalves, António Alves da Silva, et al. "Natural Woodlands Hold More Diverse, Abundant, and Unique Biota Than Novel Anthropogenic Forests: A Multi-group Assessment." *European Journal of Forest Research* 138, no. 3 (2019): 461–72; Fork, Susanne, and Vânia M. Proença. "Biodiversity Effects and Rates of Spread of Nonnative Eucalypt Woodlands in Central California." *Ecological Applications* 25, no. 8 (2015): 2306–19.

6. public support for reforestation efforts . . . Peterson St-Laurent, G., S. Hagerman, and R. Kozak. "What Risks Matter? Public Views About Assisted Migration and Other Climate-Adaptive Reforestation Strategies." *Climatic Change* 151, no. 3 (2018): 573–87; Peterson St-Laurent, G., L. E. Oakes, M. Cross, and S. Hagerman. "R-R-T (Resistance-Resilience-Transformation) Typology Reveals Differential Conservation Approaches Across Ecosystems And Time." *Communications Biology* 4, no. 1 (2021): 1–9.

7. a study combining anthropological research . . . Rossetto, M., E. J. Ens, T. Honings, P. D. Wilson, J. Y. S. Yap, O. Costello, and C. Bowern. "From Songlines to Genomes: Prehistoric Assisted Migration of a Rain Forest Tree by Australian Aboriginal People." *PloS One* 12, no. 11 (2017): 0186663.

8. populations of limber pine . . . Vogan, P. J., and A. W. Schoettle. "Selection for Resistance to White Pine Blister Rust Affects the Abiotic Stress Tolerances of Limber Pine." *Forest Ecology and Management* 344 (2015): 110–19; Selecting for those more tolerant . . . Lauder, J. D., E. V. Moran, and S. C. Hart. "Fight or Flight? Potential Tradeoffs Between Drought Defense and Reproduction in Conifers." *Tree Physiology* 39, no. 7 (2019): 1071–85; Huang, J., M. Kautz, A. M. Trowbridge, A. Hammerbacher, K. F. Raffa, H. D. Adams, and H. Hartmann. "Tree Defence and Bark Beetles in a Drying World: Carbon Partitioning, Functioning and Modelling." *New Phytologist* 225, no. 1 (2020): 26–36; Siemens, D. H., J. Duvall-Jisha, J. Jacobs, J. Manthey, R. Haugen, and S. Matzner. "Water Deficiency Induces Evolutionary Tradeoff Between Stress Tolerance and Chemical Defense Allocation That May Help Explain Range Limits in Plants." *Oikos* 121, no. 5 (2012): 790–800; when the weather became even wetter . . . Woods, A., K. D. Coates, and A. Hamann. "Is an Unprecedented Dothistroma Needle Blight Epidemic Related to Climate Change?" *BioScience* 55, no. 9 (2005): 761–69.

9. I'd learned about giant sequoias . . . "Giant Sequoia Thrives in B.C's New Climate." Video posted to YouTube by Vancouver Sun, March 1, 2018. www.youtube.com/watch?v=vOKBdBroEQg; Some of the tallest giant sequoias . . . "Benmore Botanic Garden, Scotland." Monumental Trees. Accessed December 19, 2023. www.monumentaltrees.com/en/content/benmore; "Japan." Giant Sequoia Nursery. www.giant-sequoia.com/gallery/japan.

10. The 2020 Castle Fire burned . . . "Wildfires Kill Unprecedented Numbers of Large Sequoia Trees." US National Park Service. www.nps.gov/articles/000/wildfires-kill-unprecedented-numbers-of-large-sequoia-trees.htm; The 2022 Washburn fire . . . Chung, Christine. "Washburn Fire Threatens Sequoias in Yosemite National Park." *New York Times*, July 10, 2022. www.nytimes.com/2022/07/10/us/washburn-fire-yosemite.html.

11. planting trees isn't an activity . . . Winks, Robin W. "The National Park Service Act of 1916: A Contradictory Mandate." *Denver Law Review* [Symposium—The National Park System] 74, no. 3 (1997): 575–623.

12. a report from one of the many . . . Nydick, K., C. Brigham, and G. Bradshaw. *A Climate-Smart Resource Stewardship Strategy for Sequoia and Kings Canyon National Parks* (Washington, DC: National Park Service, US Department of the Interior, 2017).

13. studies had shown genetically distinct . . . DeSilva, R., and R. S. Dodd. "Fragmented and Isolated: Limited Gene Flow Coupled with Weak Isolation by Environment in the Paleoendemic Giant Sequoia (*Sequoiadendron giganteum*)." *American Journal of Botany* 107, no. 1 (2020): 45–55; Dodd, Richard S., and Rainbow DeSilva. "Long-Term Demographic Decline and Late Glacial Divergence in a Californian Paleoendemic: *Sequoiadendron giganteum* (Giant Sequoia)." *Ecology and Evolution* 6, no. 10 (2016): 3342–55. https://doi.org/10.1002/ece3.2122.

14. attracting over a million visitors . . . Lindt, John. "Sequoia Park Visits Fall as Pandemic Passes." *Sun-Gazette Newspaper*, June 17, 2022. https://thesungazette.com/article/news/2022/06/17/sequoia-park-visits-fall-as-pandemic-passes.

15. In the United Kingdom, foresters are moving . . . Whittet, Richard, Stephen Cavers, Richard Ennos, and Joan Cotrell. "Genetic Considerations for Provenance Choice of Native Trees Under Climate Change in England." Forestry Commission Research Report (Edinburgh: Forestry Commission, 2019). https://cdn.forestresearch.gov.uk/2019/03/fcrp030.pdf; researchers have developed an online platform . . . "Climate-Adapted Seed Tool." Reforestation Tools. Accessed December 19, 2023. https://reforestationtools.org/climate-adapted-seed-tool.

16. common garden experiments . . . Grady, K. C., S. M. Ferrier, T. E. Kolb, S. C. Hart, G. J. Allan, and T. G. Whitham. "Genetic Variation in Productivity of Foundation Riparian Species at the Edge of Their

Distribution: Implications for Restoration and Assisted Migration in a Warming Climate." *Global Change Biology* 17, no. 12 (2011): 3724–35.

Chapter 7: Expectations

1. In 2012, an elderly woman . . . GlobalPost. "Botched 'Ecco Homo' Jesus Painter, Cecilia Gimenez, Gets Last Laugh." *The World*, August 13, 2013. https://theworld.org/stories/2013-08-13/botched-ecco-homo-jesus-painter-cecilia-gimenez-gets-last-laugh.

2. Giménez claimed she hadn't finished . . . Benedictus, Leo. "Life After a Viral Nightmare: From Ecce Homo to Revenge Porn." *The Guardian*, January 7, 2015, sec. Technology. www.theguardian.com/global/2015/jan/07/life-after-a-viral-nightmare-ecce-homo-to-revenge-porn; The portrait had been transformed . . . Lonely Planet, *Secret Marvels of the World: 360 Extraordinary Places You Never Knew Existed and Where to Find Them* (Oakland, CA: Lonely Planet Publications, 2017); Parodies showed the original . . . GlobalPost. "Botched 'Ecco Homo' Jesus Painter, Cecilia Gimenez, Gets Last Laugh." *The World*, August 13, 2013. https://theworld.org/stories/2013-08-13/botched-ecco-homo-jesus-painter-cecilia-gimenez-gets-last-laugh; its monkeylike appearance . . . Lampen, Claire. "Potato Jesus Resurrects Tourism, Provides for the Elderly." *The Cut*, December 28, 2018. www.thecut.com/2018/12/botched-ecce-homo-restoration-spawns-meme-revives-tourism.html; A Facebook group titled . . . "Potato Jesus." Know Your Meme, August 2012. https://knowyourmeme.com/memes/potato-jesus.

3. a mention of *Ecce Mono* . . . Rohwer, Y., and E. Marris. "Renaming Restoration: Conceptualizing and Justifying the Activity as a Restoration of Lost Moral Value Rather Than a Return to a Previous State." *Restoration Ecology* 24, no. 5 (2016): 674–79.

4. first known economic miracle . . . Lampen, Claire. "Potato Jesus Resurrects Tourism, Provides for the Elderly." *The Cut*, December 28, 2018. www.thecut.com/2018/12/botched-ecce-homo-restoration-spawns-meme-revives-tourism.html; It had resurrected tourism . . . Carvajal, Doreen. "A Town, if Not a Painting, Is Restored." *New York Times*, December 14, 2014. www.nytimes.com/2014/12/15/world/a-town-if-not-a-painting-is-restored.html.

5. Suzanne Prober, an ecologist with . . . Prober, S. M., V. A. Doerr, L. M. Broadhurst, K. J. Williams, and F. Dickson. "Shifting the Conservation Paradigm: A Synthesis of Options for Renovating Nature Under Climate Change." *Ecological Monographs* 89, no. 1 (2019): 01333.

Chapter 8: Coexisting with Concrete

1. than seen at Yosemite . . . "Yosemite National Park Visitors 2022." Statista. Accessed December 19, 2023. www.statista.com/statistics/254232/number-of-visitors-to-the-yosemite-national-park-in-the-us; nearly 85 percent of people live . . . Center for Sustainable Systems. "U.S. Cities Factsheet." University of Michigan, 2021; Globally, about 4.4 billion people . . . "Urban Development." World Bank, April 3, 2023. www.worldbank.org/en/topic/urbandevelopment/overview.

2. Relative to nonurban areas . . . Peng, Shushi, Shilong Piao, Philippe Ciais, Pierre Friedlingstein, Catherine Ottle, François-Marie Bréon, Huijuan Nan, et al. "Surface Urban Heat Island Across 419 Global Big Cities." *Environmental Science & Technology* 46, no. 2 (January 17, 2012): 696–703. https://doi.org/10.1021/es2030438; People in urban areas . . . Akbari, H., M. Pomerantz, H. Taha, Y. Chen, N. H. Wong, T. Kroeger, R. I. McDonald, T. Boucher, et al. "Cool Surfaces and Shade Trees to Reduce Energy Use and Improve Air Quality in Urban Areas." *Solar Energy* 70, no. 3 (2001): 295–310; Yu, Chen, and Wong Nyuk Hien. "Thermal Benefits of City Parks." *Energy and Buildings* 38, no. 2 (February 1, 2006): 105–20. https://doi.org/10.1016/j.enbuild.2005.04.003; Kroeger, Timm, Robert I. McDonald, Timothy Boucher, Ping Zhang, and Longzhu Wang. "Where the People Are: Current Trends and Future Potential Targeted Investments in Urban Trees for PM10 and Temperature Mitigation in 27 U.S. Cities." *Landscape and Urban Planning* 177 (September 1, 2018): 227–40. https://doi.org/10.1016/j.landurbplan.2018.05.014; Trees growing in these urban heat islands . . . Han, Qiyao, Greg Keeffe, Paul Caplat, and Alan Simson. "Cities as Hot Stepping Stones for Tree Migration." *npj Urban Sustainability* 1, no. 1 (2021): 1–5.

3. a 10 percent increase in tree canopy cover . . . Schusler, T., L. Weiss, D. Treering, and E. Balderama. "Research Note: Examining the Association Between Tree Canopy, Parks and Crime in Chicago." *Landscape and Urban Planning* 170 (2018): 309–13.

4. a first-generation college student . . . Wilson, Beattra. "Beattra Wilson's Steadfast Path: An Urban Forestry and USDA Forest Service Journey." New York State Urban Forestry Council (blog), October 30, 2020. https://nysufc.org/beattra-wilsons-steadfast-path-an-urban-forestry-usda-forest-service-journey/2020/10/30.

5. oldest national conservation organization . . . "American Forests Announces Expansion of Transformative Tree Equity Score Tool." American Forests, November 17, 2022. www.americanforests.org/article/american-forests-announces-expansion-of-transformative-tree-equity-score-tool.

6. a legacy of redlining . . . Jackson, Candace. "What Is Redlining?" *New York Times*, August 17, 2021, sec. Real Estate. www.nytimes.com/2021/08/17/realestate/what-is-redlining.html; A study of urban tree canopy cover . . . Locke, D. H., B. Hall, J. M. Grove, S. T. Pickett, L. A. Ogden, C. Aoki, and J. P. O'Neil-Dunne. "Residential Housing Segregation and Urban Tree Canopy in 37 US Cities." *npj Urban Sustainability* 1, no. 1 (2021): 1–9.

7. Chris and his team . . . "Tree Equity Score," www.treeequityscore.org.

8. A study in Tampa, Florida . . . Jennings, Viniece. "Structural Characteristics of Tree Cover and the Association with Cardiovascular and Respiratory Health in Tampa, FL." *Journal of Urban Health* 96 (2019): 669–81; relationship between tree cover and depression . . . Browning, Matthew H. E. M., Kangjae Lee, and Kathleen L. Wolf. "Tree Cover Shows an Inverse Relationship with Depressive Symptoms in Elderly Residents Living in US Nursing Homes." *Urban Forestry and Urban Greening* 41 (2019): 23–32; an association between tree and shrub cover . . . Tallis, H., G. N. Bratman, J. F. Samhouri, and J. Fargione. "Are California Elementary School Test Scores More Strongly Associated with Urban Trees Than Poverty?" *Frontiers in Psychology* 9 (2018).

9. A study in Detroit, Michigan . . . Carmichael, Christine E., and Maureen H. McDonough. "Community Stories: Explaining Resistance to Street Tree-Planting Programs in Detroit, Michigan, USA." *Society and Natural Resources* 32, no. 5 (2019): 588–605.

10. A Bloomberg article mentioned . . . Mock, Brentin. "When People Resist City Tree-Planting, It's Not from a Lack of Awareness." Bloomberg.com, January 11, 2019. www.bloomberg.com/news/articles/2019-01-11/why-detroiters-didn-t-trust-city-tree-planting-efforts; Taylor, Dorceta E. "The State of Diversity in Environmental Organizations." Green 2.0, July 2014. https://diversegreen.org/wp-content/uploads/2021/01/FullReport_Green2.0_FINAL.pdf.

11. In Phoenix, the city council passed . . . "Minutes of the City Council Formal Meeting." City of Phoenix, April 21, 2021. www.phoenix.gov/cityclerksite/City%20Council%20Meeting%20Files/4-21-21%20Formal%20Agenda-FINAL.pdf; "Groundwork RI Leads Tree Equity Planting in Pawtucket/Central Falls." Groundwork Rhode Island, 2021. https://groundworkri.org/groundwork-ri-leads-tree-equity-planting-in-pawtucket-central-falls.

12. For 99.9 percent of humanity's existence . . . Song, C., H. Ikei, and Y. Miyazaki. "Physiological Effects of Nature Therapy: A Review of the Research in Japan." *International Journal of Environmental Research and Public Health* 13, no. 8 (2016): 781. https://doi.org/10.3390/ijerph13080781.

13. Using the dose-response model . . . Bratman, G. N., J. P. Hamilton, K. S. Hahn, G. C. Daily, and J. J. Gross. "Nature Experience Reduces Rumination and Subgenual Prefrontal Cortex Activation." *Proceedings of the National Academy of Sciences* 112, no. 28 (2015): 8567–72.

14. having a window *view* of nature . . . Ulrich, R. S. "View Through a Window May Influence Recovery from Surgery." *Science* 224, no. 4647 (1984): 420–21; he is still widely recognized . . . Ulrich, R. S. "Natural Versus Urban Scenes: Some Psychophysiological Effects." *Environment and Behavior* 13 (1981): 523–56; help moderate our mental state . . . Bratman, Gregory N., J. Paul Hamilton, and Gretchen C. Daily. "The Impacts of Nature Experience on Human Cognitive Function and Mental Health." *Annals of the New York Academy of Sciences* 1249, no. 1 (2012): 118–36.

15. regulate their exposure to terpenes . . . Tabackman, Lia. "What Are Terpenes? How These Cannabis Compounds May Influence Your High." *Insider*, August 14, 2021. www.insider.com/guides/health/terpenes; Georgian, Destinney, Niveditha Ramadoss, Chathu Dona, and Chhandak Basu. "Therapeutic and Medicinal Uses of Terpenes." *Medicinal Plants: From Farm to Pharmacy* (November 2019): 333–59.

16. Studies in this arena . . . Liu, Qiaohui, Xiaoping Wang, Jinglan Liu, Guolin Zhang, Congying An, Yuqi Liu, Xiaoli Fan, et al. "The Relationship Between the Restorative Perception of the Environment and the Physiological and Psychological Effects of Different Types of Forests on University Students."

International Journal of Environmental Research and Public Health 18, no. 22 (November 21, 2021): 122–24. https://doi.org/10.3390/ijerph182212224; a project in Tasmania . . . NHER. "'We Planted a Forest!'—The Mental Health Benefits of Ecological Restoration: A Pilot Study." *Natural History of Ecological Restoration*, October 21, 2022. https://mbgecologicalrestoration.wordpress.com/2022/10/21/we-planted-a-forest-the-mental-health-benefits-of-ecological-restoration-a-pilot-study.

17. high-pollutant areas . . . Ward, Ken, Jr. "How Black Communities Become 'Sacrifice Zones' for Industrial Air Pollution." *ProPublica*, December 21, 2021. www.propublica.org/article/how-black-communities-become-sacrifice-zones-for-industrial-air-pollution.

18. The first sawmill in Tacoma . . . Magden, Ronald. "Port of Tacoma—Thumbnail History, Part 1." HistoryLink.org, April 17, 2008. www.historylink.org/File/8592; air pollution from the smelter . . . "Tacoma Smelter." Washington State Department of Ecology. Accessed December 19, 2023. https://ecology.wa.gov/Spills-Cleanup/Contamination-cleanup/Cleanup-sites/Tacoma-smelter.

19. historically underserved and redlined . . . Tacoma Community Forestry has an online mapping system that details historic inequality through redlining, temperature, and urban heat island effect, as well as tree equity and opportunity. This can be accessed online at "Tacoma Community Forestry." ArcGIS StoryMaps, July 7, 2023. https://storymaps.arcgis.com/stories/0b0e009ae2bf4fc3850161bfdfce5740.

20. Yet the 500 million trees . . . "American Forests Announces Expansion of Transformative Tree Equity Score Tool." American Forests, November 17, 2022. www.americanforests.org/article/american-forests-announces-expansion-of-transformative-tree-equity-score-tool. Note: A 2021 calculation reported an equivalency of ninety-two million cars on an annual basis, which was picked up in the media. Staff at American Forests later corrected that estimate.

Chapter 9: Tree Time

1. The article was about Yishan Wong . . . Persio, Sofia Lotto. "Former Reddit CEO's New Startup Terraformation Raises $30 Million to Restore Forests and Tackle Climate Change." *Forbes*. Accessed December 20, 2023. www.forbes.com/sites/sofialottopersio/2021/06/09/former-reddit-ceos-new-startup-terraformation-raises-30-million-to-restore-forests-and-tackle-climate-change.

2. Yishan's very simple math . . . Wong, Yishan. "A Massive Global Reforestation Project Is How We Fix Climate Change." *Medium*, May 27, 2020. https://medium.com/@yishan/a-massive-global-reforestation-project-is-how-we-fix-climate-change-36afc6d4dc2.

3. The islands are known . . . Schuler, Timothy A. "Hawaii's Species: Endangered and Underfunded." *Hawaii Business Magazine*, October 11, 2017. www.hawaiibusiness.com/endangered-and-underfunded; Some 400 species . . . "Pacific Islands Fish and Wildlife Office | Species." US Fish and Wildlife Service. Accessed December 20, 2023. www.fws.gov/office/pacific-islands-fish-and-wildlife/species; "Endangered Species in Hawaii." Ballotpedia. Accessed December 20, 2023. https://ballotpedia.org/Endangered_species_in_Hawaii.

4. Except for the hoary bat . . . Olson, Steve. "Alien Species Pose a Severe Threat to Hawaii." In *Evolution in Hawaii: A Supplement to "Teaching About Evolution and the Nature of Science,"* 25–48 (Washington, DC: National Academies Press, 2004). https://doi.org/10.17226/10865; Strawberry guava from Brazil . . . Johnson, Tracy. "Biocontrol of Strawberry Guava." US Department of Agriculture Forest Service, 2012. www.fs.usda.gov/research/news/highlights/biocontrol-strawberry-guava.

5. a brief overview of Hawaiian language . . . "Kaona." Wehewehe Wikiwiki Hawaiian Language Dictionaries. Accessed December 20, 2023. https://hilo.hawaii.edu/wehe/?q=kaona.

6. possibility of Earth becoming uninhabitable . . . Wallace-Wells, David. "When Will the Planet Be Too Hot for Humans? Much, Much Sooner Than You Imagine." *Intelligencer*, July 9, 2017. https://nymag.com/intelligencer/2017/07/climate-change-earth-too-hot-for-humans.html.

7. guarded by the *ʻaoa* . . . "ʻiliahi." Wehewehe Wikiwiki Hawaiian Language Dictionaries. Accessed December 20, 2023. https://hilo.hawaii.edu/wehe/?q=%CA%BBiliahi.

8. words like *evapotranspiration* . . . "Piʻikomoāea." Wehewehe Wikiwiki Hawaiian Language Dictionaries. Accessed December 20, 2023. https://hilo.hawaii.edu/wehe/?q=pi%CA%BBikomo%C4%81ea.

9. The sandalwood was the first . . . Pili, Kamaka. "Aloha Authentic: History of Sandalwood in Hawaiʻi." *KHON2*, January 27, 2022. www.khon2.com/aloha-authentic/aloha-authentic-history-of-sandalwood-in-hawaii.

10. drier with climate change . . . "State and Private Forestry Fact Sheet." US Drug Agency, December 18, 2023. https://apps.fs.usda.gov/nicportal/temppdf/sfs/naweb/HI_std.pdf.

11. possible approaches for reforesting deserts . . . Wong, Yishan. "A Massive Global Reforestation Project Is How We Fix Climate Change." *Medium*, May 27, 2020. https://medium.com/@yishan/a-massive-global-reforestation-project-is-how-we-fix-climate-change-36afc6d4dc2.

12. He coauthored a viewpoint . . . Mora Rollo, A., A. Rollo, and C. Mora. "The Tree-Lined Path to Carbon Neutrality." *Nature Reviews Earth & Environment* 1, no. 7 (2020): 332; described a set of weighing scales . . . "Camilo Mora." Blue Planet—We Are 100, July 9, 2019. https://weare100.org/camilo-mora.

13. a response that I'd read in 2020 . . . Chazdon, Robin, and Pedro Brancalion. "Restoring Forests as a Means to Many Ends." *Science* 365, no. 6448 (2019): 24–25.

Chapter 10: Fairy Tales

1. "tree books" . . . Nivola, Claire A. *Planting the Trees of Kenya: The Story of Wangari Maathai* (New York: Farrar, Straus and Giroux, 2008).

2. signed the prior year . . . "Glasgow Leaders' Declaration on Forests and Land Use—UN Climate Change Conference (COP26) at the SEC—Glasgow 2021." National Archives, November 2, 2021. https://webarchive.nationalarchives.gov.uk/ukgwa/20230418175226/https://ukcop26.org/glasgow-leaders-declaration-on-forests-and-land-use.

3. The FCLP summit began . . . "Forests and Climate Leaders Partnership (FCLP) Summit." UNFCC, November 7, 2022. https://unfccc.int/event/forests-and-climate-leaders-partnership-fclp-summit.

4. Press releases and posts . . . "Forest and Climate Leaders' Partnership Launched at COP27." Nature-Based Solutions Initiative, November 8, 2022. www.naturebasedsolutionsinitiative.org/news/forest-and-climate-leaders-partnership-launched-at-cop27; "STATEMENT: Forest and Climate Leaders' Partnership and Forest-Related Funding at COP27." World Resources Institute, November 7, 2022. www.wri.org/news/statement-forest-and-climate-leaders-partnership-and-forest-related-funding-cop27.

5. At the launch, the presidents of . . . "Forests and Climate Leaders Partnership (FCLP) Summit." UNFCC, November 7, 2022. https://unfccc.int/event/forests-and-climate-leaders-partnership-fclp-summit.

6. published right before the conference . . . Dooley, K., H. Keith, A. Larson, G. Catacora-Vargas, W. Carton, K. L. Christiansen, Baa O. Enokenwa, et al. *The Land Gap Report 2022*. ResearchGate, November 2022. https://landgap.org/2022/report.

7. recalled a formula on land scarcity . . . Lambin, Eric F., and Patrick Meyfroidt. "Global Land Use Change, Economic Globalization, and the Looming Land Scarcity." *Proceedings of the National Academy of Sciences* 108, no. 9 (2011): 3465–72.

8. The constraints highlight the pressing question . . . For a thorough discussion of constraints to land use and related points, see also Meyfroidt, Patrick. "Ten Facts About Land Systems for Sustainability." *Proceedings of the National Academy of Sciences* 119, no. 7 (2022): 2109217118.

9. As ecologists have written . . . Chazdon, Robin L., Pedro H. S. Brancalion, Lars Laestadius, Aoife Bennett-Curry, Kathleen Buckingham, Chetan Kumar, Julian Moll-Rocek, et al. "When Is a Forest a Forest? Forest Concepts and Definitions in the Era of Forest and Landscape Restoration." *Ambio* 45, no. 5 (2016): 538–50.

10. SER is focused on not just . . . Food and Agriculture Organization and the International Union for Conservation of Nature Commission on Ecosystem Management. *Principles for Ecosystem Restoration to Guide the United Nations Decade 2021–2030* (Rome: FAO, 2021). www.fao.org/documents/card/en/c/CB6591EN.

11. Rebecca Lloyd, who studied how forest ecosystems . . . Lloyd, R. A., K. A. Lohse, and T. P. A. Ferré. "Influence of Road Reclamation Techniques on Forest Ecosystem Recovery." *Frontiers in Ecology and the Environment* 11, no. 2 (2013): 75–81.

12. SER created an ecological recovery wheel, . . . "A Tool for Assessing Ecosystem Recovery: The 5-Star Recovery System in Action." Society for Ecological Restoration. www.ser.org/page/SERNews3113.

13. "We started with Vietnam" . . . "A Conversation with Eric Lambin." Video posted to YouTube by Stanford Woods Institute for the Environment, March 6, 2020. www.youtube.com/watch?v=MJt1P6wRo2k.

14. Vietnam's reversal and what had caused it . . . Meyfroidt, Patrick, and Eric F. Lambin. "The Causes of the Reforestation in Vietnam." *Land Use Policy* 25, no. 2 (2008): 182–97.

15. plantations were a big part . . . Meyfroidt, Patrick, and Eric F. Lambin. "Forest Transition in Vietnam and Its Environmental Impacts." *Global Change Biology* 14, no. 6 (2008): 1319–36.

16. referring to *tree*-cover transitions instead of *forest*-cover . . . Minang, P. A., M. Noordwijk, and E. Kahurani, eds. *Partnership in the Tropical Forest Margins: A 20-Year Journey in Search of Alternatives to Slash-and-Burn* (Nairobi: World Agroforestry Centre, 2014).

17. came across a headline . . . "Vietnam's Wooden Furniture Sales to U.S. Surge." *Thanh Nien News*, May 11, 2005. https://vietnamembassy-usa.org/relations/vietnams-wooden-furniture-sales-us-surge.

18. The phenomenon is called *land-use leakage* . . . Meyfroidt, Patrick, Jan Börner, Rachael Garrett, Toby Gardner, Javier Godar, K Kis-Katos, Britaldo Filho, et al. "Focus on Leakage and Spillovers: Informing Land-Use Governance in a Tele-coupled World." *Environmental Research Letters* 15 (September 1, 2020). https://doi.org/10.1088/1748-9326/ab7397.

19. went out to investigate leakage . . . Meyfroidt, Patrick, Thomas K. Rudel, and Eric F. Lambin. "Forest Transitions, Trade, and the Global Displacement of Land Use." *Proceedings of the National Academy of Sciences* 107, no. 49 (2010): 20917–22.

20. for Bhutan to get its wood . . . Jadin, I., Patrick Meyfroidt, and Eric F. Lambin. "Forest Protection and Economic Development by Offshoring Wood Extraction: Bhutan's Clean Development Path." *Regional Environmental Change* 16 (2016): 401–15.

21. Costa Rica is another example . . . Jadin, I., Patrick Meyfroidt, and Eric F. Lambin. "International Trade, and Land Use Intensification and Spatial Reorganization Explain Costa Rica's Forest Transition." *Environmental Research Letters* 11, no. 3 (2016): 035005; scientists call this *benign leakage* . . . Meyfroidt, Patrick, Jan Börner, Rachael Garrett, Toby Gardner, Javier Godar, K. Kis-Katos, Britaldo Filho, et al. "Focus on Leakage and Spillovers: Informing Land-Use Governance in a Tele-coupled World." *Environmental Research Letters* 15 (September 1, 2020). https://doi.org/10.1088/1748-9326/ab7397; Patrick prefers *spillover* . . . Meyfroidt, Patrick, R. R. Chowdhury, A. Bremond, E. C. Ellis, K. H. Erb, T. Filatova, and P. H. Verburg. "Middle-Range Theories of Land System Change." *Global Environmental Change* 53 (2018): 52–67.

22. hinges on a widespread revolution . . . Balmford, Andrew. "Concentrating vs. Spreading Our Footprint: How to Meet Humanity's Needs at Least Cost to Nature." *Journal of Zoology* 315, no. 2 (2021): 79–109. https://doi.org/10.1111/jzo.12920.

23. Maathai, Wangari. *Replenishing the Earth: Spiritual Values for Healing Ourselves and the World* (New York: Doubleday, 2010), 14.

24. achieved through grazing rotations . . . Gewin, Virginia. "A New Study on Regenerative Grazing Complicates Climate Optimism." *Civil Eats*, January 6, 2021. https://civileats.com/2021/01/06/a-new-study-on-regenerative-grazing-complicates-climate-optimism.

25. getting soils globally to just 3 percent . . . "Indigo Launches The Terraton InitiativeTM to Remove One Trillion Tons of Carbon Dioxide from the Atmosphere by Unlocking the Potential of Agricultural Soils to Sequester Carbon." *Businesswire*, June 12, 2019. www.businesswire.com/news/home/20190612005271/en/Indigo-Launches-The-Terraton-Initiative%E2%84%A2-to-Remove-One-Trillion-Tons-of-Carbon-Dioxide-from-the-Atmosphere-by-Unlocking-the-Potential-of-Agricultural-Soils-to-Sequester-Carbon; Keenor, Sam G., Aline F. Rodrigues, Li Mao, Agnieszka E. Latawiec, Amii R. Harwood, and Brian J. Reid. "Capturing a Soil Carbon Economy." *Royal Society Open Science* 8, no. 4 (April 14, 2021): 202305. https://doi.org/10.1098/rsos.202305.

26. natural areas store more carbon . . . Williams, David R., Ben Phalan, Claire Feniuk, Rhys E. Green, and Andrew Balmford. "Carbon Storage and Land-Use Strategies in Agricultural Landscapes Across Three Continents." *Current Biology* 28, no. 15 (2018): 2500–505.

27. 200 million hectares of maize . . . "Acreage of Grain Worldwide by Type." Statista, 2022. www.statista.com/statistics/272536/acreage-of-grain-worldwide-by-type; The rapid expansion of soy farms . . . Nepstad, D. C., C. M. Stickler, O. T. Almeida, and S. Sauer. "Globalization of the Amazon Soy and Beef Industries: Opportunities for Conservation." *Conservation Biology* 20, no. 6 (2006): 1595–603; Sauer, Sérgio. "Soy Expansion into the Agricultural Frontiers of the Brazilian Amazon: The Agribusiness Economy and Its Social and Environmental Conflicts." *Land Use Policy* 79 (December 1, 2018): 326–38. https://doi.org/10.1016/j.landusepol.2018.08.030.

Chapter 11: All Hands on Deck

1. planting a trillion trees . . . "Press Conference: One Trillion Trees | DAVOS 2020." Video posted to YouTube by World Economic Forum, January 22, 2020. www.youtube.com/watch?v=8kPMtDiiXxk.

2. citywide goal of increasing tree cover . . . Elwell, Jona. "DC's Urban Forestry Division by the Numbers." *Casey Trees*, November 1, 2021. https://caseytrees.org/2021/11/dcs-urban-forestry-division-by-the-numbers.

3. another study that Tom had led . . . Crowther, Thomas W., H. B. Glick, D. S. Maynard, W. Ashley-Cantello, T. Evans, and D. Routh. "Predicting Global Forest Reforestation Potential." *bioRxiv* (2017): 210062. https://doi.org/10.1101/210062.

4. protecting and restoring wild animals . . . Schmitz, Oswald J., Magnus Sylvén, Trisha B. Atwood, Elisabeth S. Bakker, Fabio Berzaghi, Jedediah F. Brodie, Joris P. G. M. Cromsigt, et al. "Trophic Rewilding Can Expand Natural Climate Solutions." *Nature Climate Change* 13, no. 4 (April 2023): 324–33. https://doi.org/10.1038/s41558-023-01631-6.

5. Often the result of people using smoke . . . Nkurunziza, Michel. "Four Arrested over Nyungwe Fire." *New Times*, August 1, 2022.www.newtimes.co.rw/article/201687/News/four-arrested-over-nyungwe-fire.

6. the principles are well intentioned . . . Food and Agriculture Organization and the International Union for Conservation of Nature Commission on Ecosystem Management. *Principles for Ecosystem Restoration to Guide the United Nations Decade 2021–2030* (Rome: FAO, 2021). www.fao.org/documents/card/en/c/CB6591EN; similar principles, such as focusing on landscapes . . . "Forest Landscape Restoration Principles." World Resources Institute. Accessed December 20, 2023. www.wri.org/initiatives/global-restoration-initiative/forest-landscape-restoration-principles.

7. news related to tree plantings . . . Kent, Sami. "Most of 11m Trees Planted in Turkish Project 'May Be Dead.'" *The Guardian*, January 30, 2020, sec. World News. www.theguardian.com/world/2020/jan/30/most-of-11m-trees-planted-in-turkish-project-may-be-dead; Galer, Sophia Smith. "'Greenwashing': Tree-Planting Schemes Are Just Creating Tree Cemeteries." *Vice*. Accessed December 20, 2023. www.vice.com/en/article/v7v75a/tree-planting-schemes-england; Einhorn, Catrin. "Tree Planting Is Booming. Here's How That Could Help, or Harm, the Planet." *New York Times*, March 14, 2022, sec. Climate. www.nytimes.com/2022/03/14/climate/tree-planting-reforestation-climate.html; Knuth, Hannah, and Tin Fischer. "The Fairy Tale Forest." *Zeit Online*, December 16, 2020. www.zeit.de/2020/53/plant-for-the-planet-klimaschutz-organisation-mexiko-spendengelder/komplettansicht; Pearce, Fred. "Phantom Forests: Why Ambitious Tree Planting Projects Are Failing." *Yale E360*, October 6, 2022. https://e360.yale.edu/features/phantom-forests-tree-planting-climate-change.

8. I selected other points . . . Ellis, Erle C., Mark Maslin, and Simon Lewis. "Planting Trees Won't Save the World." *New York Times*, February 12, 2020, sec. Opinion. www.nytimes.com/2020/02/12/opinion/trump-climate-change-trees.html.

9. Contacts at The Nature Conservancy pointed . . . Haugen, JoAnna. "Seed by Seed, a Women's Collective Helps Reforest Brazil's Xingu River Basin." *Mongabay Environmental News*, May 12, 2020. https://news.mongabay.com/2020/05/seed-by-seed-a-womens-collective-helps-reforest-brazils-xingu-river-basin.

10. the most destructive and costly fire . . . "Camp Fire (2018)." Wikipedia, December 12, 2023. https://en.wikipedia.org/w/index.php?title=Camp_Fire_(2018)&oldid=1189587832; strategies for assisting that regeneration . . . Rempel, Austin, Britta Dyer, Brooke Thompson, Coreen Francis, Heidi Rogers, Kristin Schmitt, Leana Weissberg, et al. "Camp Fire Restoration Plan." American Forests, 2021. www.americanforests.org/wp-content/uploads/2021/10/BLM_CampPlan_web.pdf.

11. The Great Green Wall . . . "The Great Green Wall." *National Geographic*. Accessed December 20, 2023. https://education.nationalgeographic.org/resource/great-green-wall.

12. Take the National Regreening Project of the Philippines . . . "Enhanced National Greening Program." Department of Environment and Natural Resources. Accessed December 20, 2023. www.denr.gov.ph/index.php/priority-programs/national-greening-program.

13. active in Kenya, Tanzania, Uganda, and India . . . "Plant Trees and Buy Farmers-First Carbon Credits." TIST Program. https://program.tist.org.

14. Farmers collect the seeds . . . "Farmers." TIST Program. https://program.tist.org/farmers.

15. Kevin referenced a report . . . "Nature Risk Rising: Why the Crisis Engulfing Nature Matters for Business and the Economy." World Economic Forum, 2020. http://www3.weforum.org/docs/WEF_New_Nature_Economy_ Report_2020.pdf.

16. value of services coming from nature . . . Costanza, Robert, Ralph d'Arge, Rudolf de Groot, Stephen Farber, Monica Grasso, Bruce Hannon, Karin Limburg, et al. "The Value of the World's Ecosystem Services and Natural Capital." *Nature* 387, no. 6630 (May 1997): 253–60. https://doi.org/10.1038/387253a0.

Chapter 12: Incentivize

1. In the version covered . . . *Dallas Morning News*. "Clinton to Lead Environmental Summit." *Deseret News*. Accessed December 21, 2023. www.deseret.com/1997/5/10/19311287/clinton-to-lead-environmental-summit; "Clinton Visits a Rain Forest." *AllPolitics*, May 9, 1997. https://edition.cnn.com/ALLPOLITICS/1997/05/09/clinton.trip.

2. implementing payments for ecosystem services . . . Pagiola, Stefano. "Payments for Environmental Services in Costa Rica." *Ecological Economics* 65, no. 4 (2008): 712–24.

3. Norway had purchased . . . Miranda, Miriam, Ina T Porras, and Mary Luz Moreno. "The Social Impacts of Payments for Environmental Services in Costa Rica." Environmental Economics Programme, October 2003.

4. International Monetary Fund reported a global average . . . Vitor Gaspar and Ian Parry. "A Proposal to Scale Up Global Carbon Pricing." International Monetary Fund, June 18, 2021. www.imf.org/en/Blogs/Articles/2021/06/18/blog-a-proposal-to-scale-up-global-carbon-pricing; a peak in pricing for nature-based credits . . . Kenza, Bryan. "The Looming Land Grab in Africa for Carbon Credits." *Financial Times*, December 6, 2023. https://www-ft-com.stanford.idm.oclc.org/content/f9bead69-7401-44fe-8db9-1c4063ae958c.

5. a report with Trillion Trees . . . Oakes, Lauren E., T. Rayden, J. Lotspeich, and A. Bagwill. "Defining the Real Costs of Restoring Forests: Practical Steps Towards Improving Cost Estimates." Trillion Trees, 2022. https://trilliontrees.org/wp-content/uploads/2022/08/Trillion-Trees_Defining-the-real-cost-of-restoring-forests.pdf; widespread misperception that planting . . . "A Tree Is Planted for One Euro. What Is Paid For with the Euro?" Plant-for-the-Planet. www.plant-for-the-planet.org/faq/a-tree-is-planted-for-one-euro-what-is-paid-for-with-the-euro; as low as fifteen cents . . . George, Zach St. "Can Planting a Trillion New Trees Save the World?" *New York Times*, July 13, 2022, sec. Magazine. www.nytimes.com/2022/07/13/magazine/planting-trees-climate-change.html.

6. one tree sequesters . . . Winrock-IUCN Global Emissions and Removals Database. InfoFLR https://infoflr.org; Bernal, Blanca, Lara T. Murray, and Timothy R. H. Pearson. "Global Carbon Dioxide Removal Rates from Forest Landscape Restoration Activities." *Carbon Balance and Management* 13, no. 1 (2018): 1–13.

7. of Brazil's Atlantic Forest . . . Pedro, H. S., Joannès Guillemot, Ricardo G. César, Henrique S. Andrade, Alex Mendes, Taísi B. Sorrini, Marisa D. C. Piccolo, et al. "The Cost of Restoring Carbon Stocks in Brazil's Atlantic Forest." *Land Degradation & Development* 32, no. 2 (2021): 83–41; data from Cambodia . . . Warren-Thomas, Eleanor M., David P. Edwards, Daniel P. Bebber, Phourin Chhang, Alex N. Diment, Tom D. Evans, Frances H. Lambrick, et al. "Protecting Tropical Forests from the Rapid Expansion of Rubber Using Carbon Payments." *Nature Communications* 9 (2018). https://doi.org/10.1038/s41467-018-03287-9; In a book I'd read . . . Doerr, John. *Speed and Scale: An Action Plan for Solving Our Climate Crisis Now* (New York: Penguin, 2021).

8. The *New Yorker* ran a cartoon . . . Fisher, Ed. "Gardens." *New Yorker*, October 16, 1989. https://archives-newyorker-com.stanford.idm.oclc.org/newyorker/1989-10-16/flipbook/046.

9. the media flurry surrounding . . . Blanco, Daniel Bastardo. "We Can't Just Plant Billions of Trees to Stop Climate Change." *Discover Magazine*, April 11, 2023 (first published July 10, 2019). www.discovermagazine.com/planet-earth/we-cant-just-plant-billions-of-trees-to-stop-climate-change; Fischetti, Mark. "Massive Forest Restoration Could Greatly Slow Global Warming." *Scientific American*. Accessed December 21, 2023. www.scientificamerican.com/article/massive-forest-restoration-could-greatly-slow-global-warming.

10. a short NPR story from 2015 . . . Smith, Stacey Vanek. "NPR Amazon Reporting Team Tries to Offset Its Carbon Footprint." *NPR*, November 12, 2015. www.npr.org/2015/11/12/455717415/npr-amazon-reporting-team-tries-to-offset-its-carbon-footprint.

11. one of the pioneers of ecosystem services . . . Pineda, Fernanda. "The Professor Who Assigns Value to Nature—Then Persuades World Leaders to Save It." *Washington Post*. Accessed December 21, 2023. www.washingtonpost.com/climate-solutions/interactive/2021/gretchen-daily-natural-capital-environment.

12. the online platform . . . "Global Forest Watch." World Resources Institute, November 9, 2023. www.wri.org/initiatives/global-forest-watch; "Global Forest Watch." Global Land Analysis and Discovery, 2013. https://glad.umd.edu/projects/global-forest-watch; Vizzuality. "Forest Monitoring, Land Use and Deforestation Trends." Global Forest Watch. Accessed January 8, 2024. www.globalforestwatch.org.

13. The estimate turned out to be high . . . Roger Sant explains the fifty-two million trees as a "rather large overshoot" when considering, as he wrote, "a 180-megawatt coal-fired power plant emits about eighteen million tons of CO_2 over its 30-year lifetime" and the average amount of carbon a tree sequesters over a 40-year life. According to Roger, AES "only needed to plant eighteen million trees, not fifty-two million."

14. John Kerry helped put together . . . Curwood, Steve. "The Democratic Ticket on the Environment." *Living on Earth*, PRX, June 23, 2004; The goal of the Acid Rain Program . . . H.R.3030—101st Congress (1989–1990): Clean Air Act Amendments of 1990, Title V. Acid Deposition Control. Congress.gov, May 23, 1990. www.congress.gov/bill/101st-congress/house-bill/3030; Stavins, Robert. "The U.S. Sulphur Dioxide Cap and Trade Programme and Lessons for Climate Policy." *Vox*, August 12, 2012. www.hks.harvard.edu/publications/us-sulphur-dioxide-cap-and-trade-programme-and-lessons-climate-policy.

15. Mi Cuenca . . . Zwick, Steve. "Six Lessons from the History of Natural Climate Solutions." *Ecosystem Marketplace*. Accessed December 21, 2023. www.ecosystemmarketplace.com/articles/opinionsix-lessons-from-the-history-of-natural-climate-solutions.

16. Mark Trexler, an associate in the . . . Trexler, Mark. "Time to Rethink Nature-Based Solutions?" illuminem, August 24, 2021. https://illuminem.com/illuminemvoices/time-to-rethink-nature-based-solutions.

17. AES put forward $2 million . . . Trent, Marcy. "The AES Corporation (A)." World Resources Institute. http://pdf.wri.org/bell/case_1-56973-122-5_full_version_a_english.pdf; Activities included establishing agroforestry systems . . . Wittman, Hannah K., and Cynthia Caron. "Carbon Offsets and Inequality: Social Costs and Co-benefits in Guatemala and Sri Lanka." *Society and Natural Resources* 22, no. 8 (2009): 710–26.

18. The CARE project became known . . . Faeth, Paul. "Evaluating the Carbon Sequestration Benefits of Sustainable Forestry Projects in Developing Countries." World Resources Institute, February 1, 1994. www.wri.org/evaluating-carbon-sequestration-benefits-sustainable-forestry-projects-developing-countries.

19. AES/CARE project's first ten years . . . Brown, S., and M. Delaney. "Carbon Sequestration Final Evaluation: Final Report to CARE Guatemala for PNO3 Agroforestry Project" (Arlington, VA: Winrock International, 1999); Internal CARE memos suggested . . . Wittman, Hannah K., and Cynthia Caron. "Carbon Offsets and Inequality: Social Costs and Co-benefits in Guatemala and Sri Lanka." *Society and Natural Resources* 22, no. 8 (2009): 710–26; I found media clips online . . . Lang, Chris. "How a Forestry Offset Project in Guatemala Allowed Emissions in the USA to Increase." *REDD-Monitor*, October 9, 2009. https://reddmonitor.substack.com/p/how-a-forestry-offset-project-in.

20. The United Nations has new initiatives underway . . . Eggerts, Elizabeth, and Amanda Bradley. "Listening to Women's Voices: UN-REDD Efforts to Integrate Women's Perspectives into the Voluntary Carbon Market." UNREDD Programme, September 4, 2023. www.un-redd.org/post/listening-womens-voices-un-redd-efforts-integrate-womens-perspectives-voluntary-carbon-market.

21. In a 2022 interview . . . Bloomberg. "Transcript Zero Episode 18: How Carbon Offsets Went Wrong." Energy Connects, December 1, 2022. www.energyconnects.com/news/renewables/2022/december/transcript-zero-episode-18-how-carbon-offsets-went-wrong/.

22. *Time*, on the other hand, published a short article . . . "UTILITIES: Antidote for a Smokestack." *Time*, October 24, 1988. https://content.time.com/time/subscriber/article/0,33009,968736,00.html.

23. Watershed, a company that helps corporations . . . "Accelerating Decarbonization with $70M in New Funding." Watershed, February 8, 2022. https://watershed.com/blog/series-b.

24. A 2021 report by McKinsey . . . Blaufelder, Christopher, Cindy Levy, Peter Mannion, and Dickon Pinner. "Carbon Credits: Scaling Voluntary Markets." McKinsey, January 29, 2021. www.mckinsey.com/capabilities/sustainability/our-insights/a-blueprint-for-scaling-voluntary-carbon-markets-to-meet-the-climate-challenge#.

25. a series of articles that raised alarm bells . . . Lakhani, Nina. "'Worthless': Chevron's Carbon Offsets Are Mostly Junk and Some May Harm, Research Says." *The Guardian*, May 24, 2023, sec. Environment. www.theguardian.com/environment/2023/may/24/chevron-carbon-offset-climate-crisis; Rathi, Akshat, and Ben Elgin. "What Are Carbon Offsets and How Many Really Work?" Bloomberg.com, June 14, 2022. www.bloomberg.com/news/articles/2022-06-14/what-are-carbon-offsets-and-how-many-really-work-quicktake.

26. *Insetting* . . . might offer another pathway . . . Bhatia, Vidhi. "Carbon Insetting vs Offsetting—an Explainer." World Economic Forum, March 18, 2022. www.weforum.org/agenda/2022/03/carbon-insetting-vs-offsetting-an-explainer.

Chapter 13: Guardians of Potential

1. Replicates had made it . . . "The Kilimanjaro Project: Seed Banking to Help Scale Restoration of Forests, Rainfall, and Rivers." Terraformation, April 29, 2022. https://terraformation.com/blog/new-partner-kilimanjaro-project-tanzania.

2. China became remarkably successful . . . Chen, Chi, Taejin Park, Xuhui Wang, Shilong Piao, Baodong Xu, Rajiv K. Chaturvedi, and Richard Fuchs. "China and India Lead in Greening of the World Through Land-Use Management." *Nature Sustainability* 2, no. 2 (2019): 122–29.

3. between two trillion and seventeen trillion seeds are necessary . . . Chau, Marian, Daniela Angelova, Alice Sacco, Jill Wagner, Diana Castillo-Diaz, Victoria Meyer, and Uromi Goodale. "The Global Seed Bank Index—Thousands of Seed Banks Are Needed to Address Seed Supply Shortages in Ecosystem Restoration." ResearchGate, September 2022.

4. can yield massive seed quantities for a limited number of species . . . EPIC. "State of the World's Trees." Botanic Gardens Conservation International. Accessed December 23, 2023. www.bgci.org/resources/bgci-tools-and-resources/state-of-the-worlds-trees.

5. A 2020 analysis led by Susan Cook-Patton . . . Cook-Patton, Susan C., Trisha Gopalakrishna, Adam Daigneault, Sara M. Leavitt, Jim Platt, Sabrina M. Scull, Oyut Amarjargal, et al. "Lower Cost and More Feasible Options to Restore Forest Cover in the Contiguous United States for Climate Mitigation." *One Earth* 3, no. 6 (2020): 739–52. https://doi.org/10.1016/j.oneear.2020.11.013; A 2020 survey of nurseries . . . Fargione, Joseph, Diane L. Haase, Owen T. Burney, Olga A. Kildisheva, Greg Edge, Susan C. Cook-Patton, Teresa Chapman, et al. "Challenges to the Reforestation Pipeline in the United States." *Frontiers in Forests and Global Change* 4 (2021). www.frontiersin.org/articles/10.3389/ffgc.2021.629198.

6. An analysis of the seed systems . . . Bosshard, Ennia, Riina Jalonen, Tania Kanchanarak, Vivi Yuskianti, Enrique Tolentino Jr., Rekha R. Warrier, and Smitha Krishnan. "Are Tree Seed Systems for Forest Landscape Restoration Fit for Purpose? An Analysis of Four Asian Countries." *Diversity* 13, no. 11 (2021): 575; Similar supply issues for native species . . . Shaw, Kirsty, Godfrey Ruyonga, and Mark Nicholson. "Enhancing Tree Conservation and Forest Restoration in East Africa." *BGjournal* 13, no. 2 (2016): 28–31.

7. Climate change is already impacting masting . . . Bogdziewicz, Michał. "How Will Global Change Affect Plant Reproduction? A Framework for Mast Seeding Trends." *New Phytologist* 234, no. 1 (2022): 14–20. https://doi.org/10.1111/nph.17682; stressors will continue to affect . . . Hernandez, Jonathan O., Muhammad Naeem, and Wajid Zaman. "How Does Changing Environment Influence Plant Seed Movements as Populations of Dispersal Vectors Decline?" *Plants* 12, no. 7 (2023): 14–62.

8. As a young boy, Vavilov . . . Nabhan, Gary Paul. *Where Our Food Comes From: Nickolay Vavilov's Quest to End Famine* (Washington, DC: Island Press, 2012). https://islandpress.org/books/where-our-food-comes.

9. Vavilov observed nature and noticed . . . Solberg, Svein Ø., Igor G. Loskutov, Line Breian, and Axel Diederichsen. "The Impact of N. I. Vavilov on the Conservation and Use of Plant Genetic Resources in Scandinavia: A Review." *Plants* 12, no. 1 (December 28, 2022): 143. https://doi.org/10.3390/plants12010143; Brezhnev, D. D. "Vsesojuznomu Ordena Lenina Nauchno-Issledovatel'skomu Institutu Rastenievodstva Imeni N. I. Vavilova—75 Let [The All-Union Order of Lenin Scientific Research Institute for Plant Industry Named After N. I. Vavilov Is 75 Years Old]." *Trudy po Prikladnoi Botanike, Genetike i Selektsii*, no. 41 (1969): 5–30; Dzyubenko, N. I. "Vavilov's Collection of Worldwide Crop Genetic Resources in the 21st Century." *Biopreservation and Biobanking* 16, no. 5 (October 2018): 377–83. https://doi.org/10.1089/bio.2018.0045.

10. a global network for food security... Nowakowski, Teresa. "Take a Virtual Tour of the 'Doomsday' Seed Vault." *Smithsonian Magazine*. Accessed December 23, 2023. www.smithsonianmag.com/smart-news/take-a-virtual-tour-of-the-doomsday-seed-vault-180981815; Liu, Rita. "Seed Banks: The Last Line of Defense Against a Threatening Global Food Crisis." *The Guardian*, April 15, 2022, sec. Environment. www.theguardian.com/environment/2022/apr/15/seed-banks-the-last-line-of-defense-against-a-threatening-global-food-crisis; seed banks have also become widespread . . . Falcis, E., D. Gauchan, R. Nankya, S. Martinez Cotto, D. I. Jarvis, L. Lewis, and P. Santis. "Strengthening the Economic Sustainability of Community Seed Banks: A Sustainable Approach to Enhance Agrobiodiversity in the Production Systems in Low-Income Countries." *Frontiers in Sustainable Food Systems* 6 (2022): 803195.

11. the *reforestation pipeline*... Fargione, Joseph, Diane L. Haase, Owen T. Burney, Olga A. Kildisheva, Greg Edge, Susan C. Cook-Patton, Teresa Chapman, et al. "Challenges to the Reforestation Pipeline in the United States." *Frontiers in Forests and Global Change* 4 (2021). www.frontiersin.org/articles/10.3389/ffgc.2021.629198.

12. the database is just one of the many resources... "Seed Information Database," https://ser-sid.org.

Chapter 14: Credible Lines of Attack

1. an article from *The Guardian* . . . Barkham, Patrick. "Fir's Fair: UK Must Embrace Conifers in Climate Fight, Says Forestry Chief." *The Guardian*, February 26, 2020, sec. Environment. www.theguardian.com/environment/2020/feb/26/firs-fair-uk-must-embrace-conifers-in-climate-fight-says-forestry-chief; The land use changes . . . Levy, Sharon. "Scotland's Bogs Reveal a Secret Paradise for Birds and Beetles." *The Guardian*, November 27, 2019, sec. Environment. www.theguardian.com/environment/2019/nov/27/scotlands-peat-bogs-reveal-their-secret-strength-carbon-aoe.

2. a so-called green rush . . . O'Grady, Cathleen. "Scotland's Billionaires Are Turning Climate Change into a Trophy Game." *The Atlantic*, May 20, 2022. www.theatlantic.com/science/archive/2022/05/scotland-climate-change-land-use/629835.

3. *Carbon colonialism*, typically defined . . . Parsons, Laurie, and Alesha de Fonseka. "Carbon Colonialism." *Decolonising Geography*, November 3, 2021. https://decolonisegeography.com/blog/2021/11/carbon-colonialism; Dehm, Julia. "Carbon Colonialism or Climate Justice: Interrogating the International Climate Regime from a TWAIL Perspective." *Windsor Yearbook of Access to Justice* 33, no. 3 (2016): 129–61. https://doi.org/10.22329/wyaj.v33i3.4893.

4. majority of afforestation projects pursuing credits . . . "WCC Statistics." UK Woodland Carbon Code. Accessed December 23, 2023. www.woodlandcarboncode.org.uk/uk-land-carbon-registry/wcc-statistics#area.

5. UK government is targeting a rise . . . "Net Zero—the UK's Contribution to Stopping Global Warming." Climate Change Committee, May 2, 2019. www.theccc.org.uk/publication/net-zero-the-uks-contribution-to-stopping-global-warming; government's Climate Change Committee . . . "Land Use: Reducing Emissions and Preparing for Climate Change." Climate Change Committee, November 15, 2018. www.theccc.org.uk/publication/land-use-reducing-emissions-and-preparing-for-climate-change.

6. Restored mountain woodlands in Scotland . . . Watts, Sarah H., and Alistair S. Jump. "The Benefits of Mountain Woodland Restoration." *2022* 30, no. 8 (April 10, 2022). https://onlinelibrary-wiley-com.stanford.idm.oclc.org/doi/full/10.1111/rec.13701.

7. *Pavlovnia* in honor of Anna Pavlovna . . . Basu, Chhandak, Nirmal Joshee, Tigran Gezalian, Brajesh Vaidya, Asada Satidkit, Homa Hemmati, and Zachary Perry. "Cross-Species PCR and Field Studies on Paulownia Elongata: A Potential Bioenergy Crop." *Bioethanol* 2 (December 18, 2015): 12–23. https://doi.org/10.1515/bioeth-2015-0002.

8. growing an acre of *Paulownia* . . . Chasan, Emily. "We Already Have the World's Most Efficient Carbon Capture Technology." Bloomberg.com, August 2, 2019. www.bloomberg.com/news/features/2019-08-02/we-already-have-the-world-s-most-efficient-carbon-capture-technology; studies have deemed its use in agroforestry . . . Magar, Lila, Saraswoti Khadka, Uttam Pokharel, Nabin Rana, Puspa Thapa, Uddhav Khadka, Jay Joshi, et al. "Total Biomass Carbon Sequestration Ability Under the Changing Climatic Condition by Paulownia Tomentosa Steud." *International Journal of Applied Sciences and Biotechnology* 6, no. 3 (October 1, 2018): 220–26. https://doi.org/10.3126/ijasbt.v6i3.20772; Jensen, Janus

Bojesen, "An Investigation into the Suitability of Paulownia as an Agroforestry Species for UK & NW European Farming Systems." Department of Agriculture & Business Management, SRUC, May 2016. www.researchgate.net/publication/311558333_An_investigation_into_the_suitability_of_Paulownia_as_an_agroforestry_species_for_UK_NW_European_farming_systems.

9. track below-ground carbon and nutrients . . . "TerraMap Gold—a New and Unrivalled Insight into Soils." South East Farmer, July 1, 2022. www.southeastfarmer.net/arable/terramap-gold-a-new-and-unrivalled-insight-into-soils; "TerraMap." Omnia. https://omniadigital.co.uk/our-services/terra-map.

10. BP holding the rights . . . "Scottish Forest Alliance." Markit Environmental Registry. Accessed December 23, 2023. https://mer.markit.com/br-reg/public/master-project.jsp?project_id=103000000000757; Perks, Mike, L. Nagy, M. Auld, M. Wood, N. Atkinson, L. Staples-Scott, G. Harvey, et al. "Carbon Sequestration Benefits of New Native Woodland Expansion in Scotland." Scottish Forest Alliance. Accessed December 23, 2023. https://thegreattrossachsforest.co.uk/assets/pdfs/great-forest-restored/CFC-Scottish-forest-Alliancecarbon-sequestration-paper.pdf.

11. forest protection in Cambodia and Peru . . . Murphy, Jazmin. "Warning: British Airways Only Climate Neutral for Domestic UK Flights." 8 Billion Trees: Carbon Offset Projects and Ecological Footprint Calculators, June 21, 2021. https://8billiontrees.com/carbon-offsets-credits/flights-airline-travel/british-airways.

Chapter 15: Luck Forest

1. Tropical dry forests are generally found . . . "Tropical Dry Forest | Description, Biome, Ecosystem, Plants, Animals, and Facts." *Britannica*, November 5, 2023. www.britannica.com/science/tropical-dry-forest.

2. The company had received $14.3 million . . . "Earthshot Labs—Funding, Financials, Valuation and Investors." Crunchbase. Accessed December 23, 2023. www.crunchbase.com/organization/earthshot-labs/company_financials.

3. about $49 per tonne . . . Microsoft went public with its investment in the project in May 2024, and news coverage highlighted a purchase of 1.6 million carbon removal credits. Two other companies, Carbon Streaming and Rubicon Carbon, also provided financing. At the time, the backers did not provide details to the media on the final carbon pricing. (See, e.g., Furness, Virginia. "Microsoft Backs Tree-Planting Carbon Removal Scheme in Panama." *Reuters*, May 22, 2024.)

4. Pro Eco Azuero, a local organization . . . Stirn, Matt. "In Search of Panama's Elusive Spider Monkeys." *New York Times*, February 28, 2022, sec. Travel. www.nytimes.com/2022/02/28/travel/panama-azuero-spider-monkeys.html; Hance, Jeremy. "Saving One of the Last Tropical Dry Forests, an Interview with Edwina von Gal." *Mongabay Environmental News*, June 29, 2009, sec. Environmental News. https://news.mongabay.com/2009/06/saving-one-of-the-last-tropical-dry-forests-an-interview-with-edwina-von-gal.

5. descriptions I'd reviewed reported 10,000 hectares . . . "How Big Is 25,000 Acres?" The Measure of Things. Accessed December 23, 2023. www.themeasureofthings.com/results.php?comp=area&unit=a&amt=25000&sort=pr&p=1.

6. right trees in the right places . . . DeWitt, Sean, Jared Messinger, and Nadia Peimbert-Rappaport. "Want to Grow Trees? Consider These 5 Lessons." World Resources Institute, February 10, 2020. www.wri.org/insights/want-grow-trees-consider-these-5-lessons; Di Sacco, Alice, Kate A. Hardwick, David Blakesley, Pedro H. S. Brancalion, Elinor Breman, Loic Cecilio Rebola, Susan Chomba, et al. "Ten Golden Rules for Reforestation to Optimize Carbon Sequestration, Biodiversity Recovery and Livelihood Benefits." *Global Change Biology* 27, no. 7 (2021): 1328–48; Bateman, Ian J., Karen Anderson, Arthur Argles, Claire Belcher, Richard A. Betts, Amy Binner, Richard E. Brazier, et al. "A Review of Planting Principles to Identify the Right Place for the Right Tree for 'Net Zero Plus' Woodlands: Applying a Place-Based Natural Capital Framework for Sustainable, Efficient and Equitable (SEE) Decisions." *People and Nature* 5, no. 2 (2023): 271–301.

7. Crowther had made a similar argument . . . Crowther, Thomas W. "Healthy Biodiversity Is the Reason to Fight Climate Change." *Time*, November 16, 2022. https://time.com/6234466/healthy-biodiversity-climate-change.

Chapter 16: With Dreams of Future Life

1. football fields' worth of tropical forest . . . Hood, Marlowe. "Football Pitch of Tropical Forest Lost Every 5 Seconds." Phys.org, June 27, 2023. https://phys.org/news/2023-06-football-pitch-tropical-forest-lost.html.

2. Bill Gates told a *New York Times* reporter . . . Gelles, David. "Billionaire Moguls and a Trillion Trees." *New York Times*, September 26, 2023, sec. Climate. www.nytimes.com/2023/09/26/climate/billionaire-moguls-and-a-trillion-trees.html.

3. benefits that come from keeping our forests . . . Harris, Nancy L., David A. Gibbs, Alessandro Baccini, Richard A. Birdsey, Sytze de Bruin, Mary Farina, Lola Fatoyinbo, et al. "Global Maps of Twenty-First Century Forest Carbon Fluxes." *Nature Climate Change* 11, no. 3 (March 2021): 234–40. https://doi.org/10.1038/s41558-020-00976-6; average amount of carbon dioxide sequestered annually . . . Friedlingstein, Pierre, Michael O'Sullivan, Matthew W. Jones, Robbie M. Andrew, Dorothee C. E. Bakker, Judith Hauck, Peter Landschützer, et al. "Global Carbon Budget 2023." *Earth System Science Data* 15, no. 12 (December 5, 2023): 5301–69. https://doi.org/10.5194/essd-15-5301-2023.

4. Indigenous forests . . . Veit, Peter, David Gibbs, and Katie Reytar. "Indigenous Forests Are Some of the Amazon's Last Carbon Sinks." World Resources Institute, January 6, 2023. www.wri.org/insights/amazon-carbon-sink-indigenous-forests; annual carbon emissions . . . "Global Carbon Atlas." Global Carbon Atlas. Accessed December 23, 2023. https://globalcarbonatlas.org.

5. study, published in *Nature* . . . Mo, Lidong, et al. "Integrated Global Assessment of the Natural Forest Carbon Potential." *Nature* 624.7990 (2023): 92–101. https://doi.org/10.1038/s41586-023-06723-z; it made the *New York Times* . . . Einhorn, Catrin, "How Much Can Trees Fight Climate Change? Massively, but Not Alone, Study Finds." *New York Times*, November 13, 2023. www.nytimes.com/2023/11/13/climate/trillion-trees-research.html; 2019 Global Potential study . . . Bastin, Jean-Francois, Yelena Finegold, Claude Garcia, Danilo Mollicone, Marcelo Rezende, Devin Routh, Constantin M. Zohner, et al. "The Global Tree Restoration Potential." *Science* 365, no. 6448 (July 5, 2019): 76–79. https://doi.org/10.1126/science.aax0848.

6. Methane, for example . . . Figure SPM.2 in "IPCC, 2021: Summary for Policymakers." In *Climate Change 2021: The Physical Science Basis. Contribution of Working Group I to the Sixth Assessment Report of the Intergovernmental Panel on Climate Change*, ed. V. Masson-Delmotte, P. Zhai, A. Pirani, S. L. Connors, C. Péan, S. Berger, N. Caud, et al. (Cambridge: Cambridge University Press, 2023), 3–32. doi: 10.1017/9781009157896.001.

7. the projected rates for annual sequestration . . . Griscom, Bronson W., Justin Adams, Peter W. Ellis, Richard A. Houghton, Guy Lomax, Daniela A. Miteva, William H. Schlesinger, et al. "Natural Climate Solutions." *Proceedings of the National Academy of Sciences* 114, no. 44 (2017): 11645–50; Figure SPM.7 in *Climate Change 2022: Mitigation of Climate Change: Working Group III Contribution to the Sixth Assessment Report of the Intergovernmental Panel on Climate Change*, ed. Priyadarshi R. Shukla, Jim Skea, Raphael Slade, Roger Fradera, Minal Pathak, Alaa Al Khourdajie, Malek Belkacemi, et al. (Geneva: Intergovernmental Panel on Climate Change, 2022). www.ipcc.ch/report/ar6/wg3/figures/summary-for-policymakers/figure-spm-7; A 2021 report . . . "The Case for Negative Emissions: A Call for Immediate Action." Coalition for Negative Emissions, 2021. https://coalitionfornegativeemissions.org/wp-content/uploads/2021/06/The-Case-for-Negative-Emissions-Coalition-for-Negative-Emissions-report-FINAL-2021-06-30.pdf.

8. carbon dioxide that can be "spent" . . . Friedlingstein, Pierre, Michael O'Sullivan, Matthew W. Jones, Robbie M. Andrew, Dorothee C. E. Bakker, Judith Hauck, Peter Landschützer, et al. "Global Carbon Budget 2023." *Earth System Science Data* 15, no. 12 (December 5, 2023): 5301–69. https://doi.org/10.5194/essd-15-5301-2023.

9. financial support for biodiversity . . . Deutz, Andrew, Geoffrey M. Heal, Rose Niu, Eric Swanson, Terry Townshend, Zhu Li, Alejandro Delmar, et al. *Financing Nature: Closing the Global Biodiveristy Financing Gap* (Paulson Institute, The Nature Conservancy, and the Cornell Atkinson Center for Sustainability, 2020); Mulder, Ivo, Aurelia Blin, Justin Adams, Teresa Hartmann, Danielle Carreira, Mark Schauer, Waltraud Ederer, et al. *State of Finance for Nature: Tripling Investments in Nature-Based Solutions by 2030* (Nairobi: United Nations Environment Programme, 2021).

10. incentivizing the private sector . . . Löfqvist, Sara, Rachael D. Garrett, and Jaboury Ghazoul. "Incentives and Barriers to Private Finance for Forest and Landscape Restoration." *Nature Ecology and Evolution* 7, no. 5 (May 2023): 707–15. https://doi.org/10.1038/s41559-023-02037-5.

11. the voluntary carbon market . . . Reuters. "Voluntary Carbon Markets Set to Become at Least Five Times Bigger by 2030—Shell." *Reuters*, January 19, 2023, sec. Carbon Markets. www.reuters.com/markets/carbon/voluntary-carbon-markets-set-become-least-five-times-bigger-by-2030-shell-2023-01-19; In 2023, nature-based projects . . . Ecosystem Marketplace. *2023 State of the Voluntary Carbon Markets* (Washington DC: Forest Trends Association, 2023).

12. The biggest polluters . . . "The Majority of People Around the World Are Concerned About Climate Change." Lloyd's Register Foundation World Risk Poll, 2019. https://wrp.lrfoundation.org.uk/2019-world-risk-poll/the-majority-of-people-around-the-world-are-concerned-about-climate-change.

13. the original sin of all forest carbon thinking . . . For more on this point, see Cullenward, D., G. Badgley, and F. Chay. "Carbon Offsets Are Incompatible with the Paris Agreement." *One Earth* 6, no. 9 (2023): 1085–88; Zickfeld, Kirsten, Deven Azevedo, Sabine Mathesius, and H. Damon Matthews. "Asymmetry in the Climate–Carbon Cycle Response to Positive and Negative CO_2 Emissions." *Nature Climate Change* 11, no. 7 (2021): 613–17; stays present in the atmosphere . . . Archer, David, Michael Eby, Victor Brovkin, Andy Ridgwell, Long Cao, Uwe Mikolajewicz, Ken Caldeira, et al. "Atmospheric Lifetime of Fossil Fuel Carbon Dioxide." *Annual Review of Earth and Planetary Sciences* 37 (2009): 117–34.

14. CO_2 emitted into the atmosphere remains . . . Joos, F., R. Roth, J. S. Fuglestvedt, G. P. Peters, I. G. Enting, W. von Bloh, V. Brovkin, et al. "Carbon Dioxide and Climate Impulse Response Functions for the Computation of Greenhouse Gas Metrics: A Multi-model Analysis." *Atmospheric Chemistry and Physics* 13, no. 5 (2013): 2793–825.

15. journalist Michael Hobbes wrote . . . Hobbes, Michael. "Stop Trying to Save the World." *New Republic*, November 17, 2014. https://newrepublic.com/article/120178/problem-international-development-and-plan-fix-it.

INDEX

Note: Page numbers with * indicate a footnote.

Credit: Clayton Boyd

LAUREN E. OAKES is a conservation scientist and science writer. She has held various appointments at Stanford University over many years, as a researcher, a lecturer, and an adjunct assistant professor in the Department of Earth System Science. Author of *In Search of the Canary Tree*, she lives in Bozeman, Montana.